The Heart

The Heart

EDITED BY JAMES PETO

YALE UNIVERSITY PRESS
NEW HAVEN AND LONDON
WELLCOME COLLECTION
LONDON

For information about this and other Yale University Press publications, please contact:
US Office: sales.press@yale.edu   www.yalebooks.com
Europe Office: sales@yaleup.co.uk   www.yaleup.co.uk

Set in Minion by J&L Composition, Filey, North Yorkshire
Printed in Italy by Conticolor

Library of Congress Cataloging-in-Publication Data

Peto, James.
   The heart/James Peto.
      p. cm.
   Includes bibliographical references.
   ISBN 978–0–300–12510–8 (cloth)
   1. Heart—Anatomy—Exhibitions.   2. Heart—Symbolic aspects—Exhibitions.   I. Title.
      QM181.P48 2007
      611'.12—dc22
                                                                      2007009220

A catalogue record for this book is available from the British Library

Contents

List of Illustrations

Introduction

Chapter 1

Chapter 2

Chapter 3

Chapter 4

Chapter 5

Foreword

Sir Magdi Yacoub

Since time immemorial, the heart has intrigued, excited and inspired human beings, and continues to do so today. As emphasised in this book, it has many facets, which arouse intense feelings, some physical, others emotional or poetic. Being, as it were, at the 'centre of things', it has extraordinary powers to influence individuals, communities, nations and beyond. And all the while, it quietly, efficiently and tirelessly supports life.

Watching the heart beating during a cardiac operation can be mesmerising; it arouses intense feelings of admiration mixed with wonder. These feelings are evoked by the heart's apparent simplicity, which is belied by its huge number of component parts. They range from molecules that cannot be seen even by the most powerful microscope, to subcellular structures measuring only a few nanometres, to larger elements like contracting muscles, blood vessels and valves, which can be observed with the naked eye. The wonder arises from the fact that these billions of widely variant components, while having their own unique and beautiful three-dimensional shape, which is perfectly suited to their function, can all act in harmony to produce the 'simple' pumping action of the heart.

Another interesting feature of the organ is that although, like any physical structure, it does conform to certain rules, each heart exhibits individual features and behaviour. This fact needs to be recognised by the operator, who should be humble, receptive and ready to adapt techniques to the individual needs of the particular heart. The operator must also be extremely patient, determined to maintain a two-way communication with the heart, and if necessary tolerate capricious behaviour from it, while trying to find a rational explanation, which is not always obvious. This harmonious relationship between the operating team and the heart, once established, should be continued during the patient's period of recovery in the intensive-care unit, avoiding frequent change of personnel, which could erode this sensitive and effective communication.

It is a remarkable fact that the vast majority of heart defects can now be mended by surgery. However, there are times when the damage to the heart is not repairable.

Until relatively recently (forty years ago), this constituted a death sentence for the patient. Heart transplantation changed that picture, with patients surviving for periods of up to thirty years, having been on the threshold of death. Unfortunately, heart transplantation is severely limited by the scarcity of donors, which necessitates the search for alternative procedures while trying to maximise donation and making the best use of available organs. Recent progress in developing and using different forms of mechanical devices, including artificial hearts and left-ventricular assist devices (LVADs) is beginning to fill a clinical need, with some of these devices replacing the function of the left ventricle or the whole heart for periods of up to three years. However, these devices cannot and will not replace all the sophisticated functions of the biological heart. One unexpected by-product of the use of LVADs, however, is the fact that a small percentage of hearts that were originally thought irreversibly damaged do actually recover, allowing eventual explantation of the device.

We at Harefield Hospital have evolved a novel combination therapy including LVADs and drugs to maximise the likelihood and durability of recovery. This form of treatment is part of a new and exciting area of medicine called 'regenerative therapy', which is being actively explored using several strategies including gene therapy and cell therapy, though these last two are strictly experimental and are not yet ready to be part of the clinical treatment of heart failure. Other even more ambitious research programmes are dedicated to tissue-engineering component parts of the heart, such as valves and heart muscle, aiming at the creation of a complete heart. This dream of engineering the heart is being actively pursued by a wide group of professionals, from clinicians, scientists, biologists and chemists to engineers and physicists.

While all of this is taking place, the heart continues to intrigue, slowly yielding up its secrets. I am confident that this book will fill an important gap in the literature, serving to shed more light on this mysterious and fascinating organ.

Introduction

James Peto

The Ancient Egyptians, when mummifying a corpse, would leave the heart inside the chest so that it could be weighed by the gods against a feather, the Symbol of Maat, or 'what is right'. This enabled them to judge the deceased's suitability for the afterlife. The brain, on the other hand, was removed from the body and discarded. Today, forty years after the first heart transplant, our understanding of where human character – and indeed life itself – is located has shifted from the heart to the brain. Yet we remain reluctant to let go of the belief that the role of the heart is somehow of much greater significance than that of a blood pump.

This book is published to accompany the exhibition *The Heart*, which opens the Wellcome Trust's new public building, Wellcome Collection. Drawing on material from everyday life and from the history of art as much as the history of science, the exhibition traces the evolution of our understanding of the heart and its anatomical and symbolic significance. The book is not intended as a catalogue of the exhibition, but rather as an opportunity to expand upon some of the ideas and issues it raises.

As well as being at the core of all things anatomical, the heart has always been central to the question of the relationship between body and soul. Reflecting the breadth of the heart's significance, each of the nine chapters approaches the subject from a very different perspective.

Louisa Young's overview and the essays by Jonathan Miller and Francis Wells show how attempts to understand the anatomical function of the heart have always been influenced by social and cultural attitudes, and that the important discoveries required the lateral thinking of figures such as Galen, Avicenna, Harvey and Leonardo da Vinci. At the same time, Fay Bound Alberti and Ayesha Nathoo demonstrate how medical breakthroughs and new anatomical knowledge can challenge our value systems and hasten social and cultural change. Examining religious imagery and rock 'n' roll respectively, the chapters by Emily Jo Sargent and Michael Bracewell underline the depth of our commitment to the idea of the heart as the seat of love and yearning. Ted

Bianco discusses the intriguing possibility of a 'universal' heartbeat throughout the animal kingdom, while Jon Turney considers the continuing struggle between the medical advances that have enabled so many hearts to beat for longer and the life-style changes that have caused others to cease beating altogether. Thanks are due to all nine authors for the diversity of viewpoints that their contributions bring.

Alternating with the nine chapters are Melissa Larner's interviews with surgeons and patients, each of whom describes a very particular relationship with his or her own heart or with the hearts of others in their professional care. The conversations they have generously contributed ensure that the book is rooted in personal experience.

The real work of putting this volume together has been done calmly and expertly by Melissa Larner. Special thanks are due to her and to Emily Jo Sargent, co-curator of the exhibition *The Heart*, and Ken Arnold, Head of Public Programmes at the Wellcome Trust, both of whom have also been instrumental in shaping its contents. Finally, I would like to thank Robert Baldock at Yale University Press, Sir Magdi Yacoub, Charmaine Griffiths at the British Heart Foundation, Rachel Hughes at Royal Brompton & Harefield NHS Trust and Francis Wells and Louisa Young for all their help and advice.

James Peto
Curator
Public Programmes
Wellcome Trust

Wellcome Collection is a free, public space devoted to exploring the links between medicine, life and art. Three galleries, events and an unrivalled library, root science in the broad context of health and well-being.

The Wellcome Trust is an independent research-funding charity, established under the will of Henry Wellcome in 1936. Its mission is to foster and promote research with the aim of improving human and animal health.

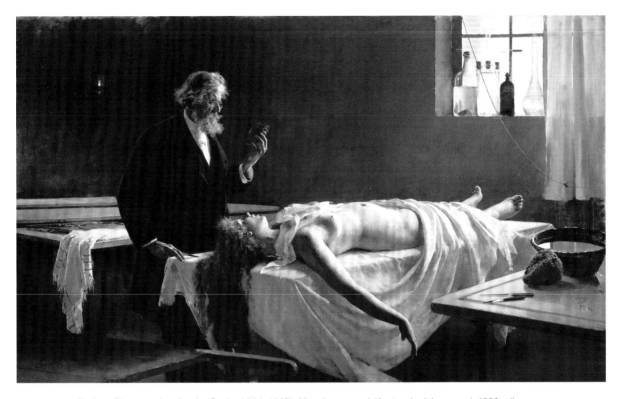

Enrique Simonet y Lombardo (Spain, 1864–1927), *Y tenia corazon! (Anatomia del corazon)*, 1890, oil on canvas, 177 × 291 cm

The Human Heart
An Overview

Louisa Young

The human heart is a piece of flesh, a lump of muscle that pumps blood around the body. So why is it often portrayed with a scalloped top and a pointy bottom, swathed in red roses with an arrow through it, and inscribed with the words 'I Love Mum', or 'Bob 4 Lou'?

The story is long and complex, and takes off in many directions, only a few of which can be followed up here. It begins in prehistoric times, and it hasn't ended yet. When I was writing *The Book of the Heart* (2002), I used to play a game: friends would think of a subject, as obscure and unlikely as possible, and I would show them how it connects to the human heart. It's not at all difficult. Everything connects to the heart.

Physically, the heart is one thing, and at the same time, it is clearly dual. It has two pairs of two different kinds of chamber. It has two types of blood – oxygenated and deoxygenated. It has two sides, sending the blood on two return journeys – to the lungs and back for oxygenation, and round the body and back, distributing oxygen. It has two types of vessel: artery and vein. Yet its purpose is unificatory: uniting the body with its nutrients, linking the body's parts, maintaining its life. This is the activity of the lump of muscle.

Long before that activity was properly understood, the heart took on another role: a spiritual and emotional purpose. The facts of the physical heart have for generations informed and reinforced the uses of the spiritual heart; here too, the heart is both double and united. The Ancient Egyptians, one of the first cultures to record their interest in the heart, and the foundation of much of our heart imagery and culture, had separate words for the heart/soul – *ib* – and the organ – *haty*.

1 Box in which to keep love letters, eighteenth century, German, papier mâché, height 20 cm

Culturally, the doubleness of the heart covers body and soul, God and man, Aristotle and Descartes, ancient and modern, sacred and profane, physical and emotional, hard and soft, male and female, recipient and creative, suction and expulsion, sex and purity, conjoined twins, red and blue, left and right, hot and cold, flesh and blood. Even its sound is double – *lubdub*. It cleaves to and from itself – 'cleave' being one of the few words that can mean one thing as well as its exact opposite: to cling together, and to cut asunder.

Because of this extraordinary versatility, the heart has been portrayed in an astounding variety of ways over the millennia. It is one of our favourite metaphors, and as such, has appeared in stories, poems, religious writings, song lyrics, paintings and sculpture, as a fruit, a flower, a seed bed, a treasure chest, a pincushion, a charm, a fountain, a house, a pump, a pine cone, a wheel, a book, a penis, a vagina, a rock, a mirror, raw material, a rose, a pomegranate, a gift, a picnic spot, a significant ingredient in pâté. It flies, it sinks, it grows, it faints, it bleeds, it flutters, it burns, it sings, it rejoices, it breaks, it fibrillates, it stops, it fails. It has an eye. It is attacked, transplanted, sacrificed, wounded, broken, thrown into the sea, given away, written on, occupied, stolen, hidden, swept, polished, eaten, filled, even circumcised: 'Circumcise yourselves to the Lord, and take away the foreskins of your heart, ye men of Judah and inhabitants of Jerusalem: lest my furie come forth like fire' (Jeremiah).

So what's it all about? Why does the heart carry this great mass of meaning? Why isn't this book about the brain, or the liver?

The simple answer is that back when we knew very little about ourselves, one thing we did know was that there was an object, right in the middle of our chests – and in the chests of the creatures we hunted and killed and ate – that moved and beat when we were alive, moved and beat faster and harder when we were excited or scared, and stopped when we were dead. As we developed an interest in our own nature, the heart became an object of great interest. Fear of death and the concommitant curiosity about what happens afterwards being perhaps the most powerful force among living creatures, the heart, by which we are either alive or dead, was a powerful entity. At some stage over the prehistoric centuries, notions of individuality, magic, religion and the life force entered our minds, bringing with them the concept of life after death, and of the soul, our immortal part. This needed to live inside the body, so where else but in the heart – this vital, central, beating, responsive organ? (There was, however, an Australian tribe that considered the soul to live in the fat of the kidneys. The Mesopotamians believed the liver to be the seat of life, and used it in divination, before it lost its status for ever under Christianity, because of this association with pagan magic and base urges.)

The earliest written mentions of the heart are emotional. The Babylonian/ Sumerian goddess Ishtar went down to the underworld to rescue her husband Tammuz from the dead. Her heart nearly broke with grief, and only joyful melodies could soothe it. Gilgamesh, hero of almost the oldest known work of literature (*The Epic of Gilgamesh*, c 2100–2000 BC), felt his heart beating with pride – he, too, was trying to bring someone back from the dead. So already, the heart is linked with love, grief, pride, courage, life, death and music.

The magical, spiritual and emotional role played by the heart contributed directly to anatomical ignorance of it: religious and mystical beliefs precluded finding out more. The hope of, desire for and belief in some form of life after death was universal among human cultures. Who could swear it would not involve the resurrection of the body? Cutting up the body to see how it worked was, for generation on generation and across the world, taboo. Anatomical knowledge therefore depended on what butchers and hunters saw inside the animals they slaughtered, and what priests saw in the animals they sacrificed, and in whose innards they fossicked for knowledge of the future. None of these was necessarily interested in anatomy; none that we know of kept a record or tried to educate others beyond passing on professional skills. The medicine men of primitive societies did sometimes open a body, but only to look for signs of magic; anatomical knowledge was not a useful thing to them.

The Ancient Egyptians, root of so much of the culture of the heart, specialised in cutting the human body. Their religious rites specifically required them to open the thoraxes of corpses in the course of mummification, and to remove certain organs. They didn't garner anatomical knowledge along the way, simply because those doing the cutting were preserving the body for eternity, not trying to uncover its secrets. The disembowelling rites, out of respect for the body, involved only tiny cuts, and the heart – which anyway was not removed during embalming – was above all for them a spiritual entity.

But they were not completely ignorant of the *haty*, the physical heart. The *Ebers Papyrus* (1550 BC), 20 metres long and written by the demigod Imhotep, contains a book that opens thus:

> The Beginning of the Secret of Medicine. Knowledge of the pulse of the heart. Knowledge of the heart . . . There are vessels for every part. In each place where each healer, each priest of Sekhmet, each magician puts his fingers – on the nape, on the hands, on the place of the heart, on the two arms, on the two feet – everywhere he encounters the heart, by its channels to all the parts.

They had identified the heart, the vessels and the pulse. Another Egyptian text, *Der Grosse Medizinische Papyrus der Berliner Museum*, speaks of 'The system of circulation of man in which is found all his maladies . . . [The vessels] bring air to his heart and it is they that give air to every part of his body.' The circulation of the blood *per se* was not discovered and proven until the English physician William Harvey (1578–1657) demonstrated it in 1618 (see Chapter 2), but it seems the Ancient Egyptians may have had an idea of it. (The Indian magico-religious text, the *Atharvaveda* of c 1200 BC also mentions circulation in the vessels, 2,800 years before Harvey). The Egyptians knew that air came in through the nose to the heart and lungs, and was circulated to all the limbs – which if you read 'oxygen' for 'air' is true, though they hadn't clarified the complexities.

The Egyptians believed that the body was fashioned by Khnum-Re, the potter god, who 'knotted the flow of the blood to the bones' and so on, all 'by the will of his heart'. The heart, they thought, slowly grew larger and heavier until the age of fifty, and then began to shrink again, just as slowly. It spoke in pulses to the organs and body parts through the *metu*, the vessels, which carried secretions, humours, blood, sperm, faeces, air, the breath of life and the breath of death. The heart governed the flow, and had a life, moods and requirements of its own. If it wandered – and it might – it was important to persuade it back into position: a heart in its place denoted good health. Even now, our heart can be in our mouth, or in our boots, neither of which gives us the same comfort and security as knowing that our heart is in the right place.

Many terms for sicknesses of the heart survive in hieroglyphs. Paragraph 855 of the Ebers Papyrus is a series of explanations of terms used to describe heart sicknesses. The heart can suffer from *wiauyt*, old age; *wegeg*, weakness; *fet*, turning aside; *maset* or *mas*, kneeling; *ad*, decay. It can be *wekh*, shrouded in darkness. It can be weary as though from travelling far, constricted, small. All these descriptions have been associated with what we now term heart failure. Congestive heart failure might be indicated by *igep*, *bah* and *meh*, meaning 'flooded' and 'drowned'. The *Ebers Papyrus* also includes what is possibly a description of angina: 'He suffers in his arm, his breast and the side of his stomach. One says concerning him: It is the *wadj* disease. Something has entered his mouth. Death is approaching.' The prognosis and symptoms are correct; *wadj* can mean 'green', arguably the skin colour of someone suffering an attack of angina. What 'enters' is usually a disease-causing demon. But then what sounds like a description of a physical heart condition might equally be an emotional or spiritual condition, or a combination of the two.

The Egyptians were themselves fairly *wekh* when it came to treatment: remedies to cool the heart were the most common; amulets were popular, in the shape of an admired animal or body part – hearts appeared frequently. Another 'timely remedy' was 'to prevent illness by having the greatness of god in your heart'.

2 Heart scarab with human face, early New Kingdom, Egyptian, green basalt, length 6.37 cm

The *ib*, the heart/soul, was of supreme importance to the Ancient Egyptians, and thus to the cultures that followed on. It is often forgotten that Islam, Christianity and Judaism all share the same God, and that only later developments – the comparative importance of prophets, and disputes about the exact parentage of one in particular – divided them. Islamic and European ideas of the heart are reassuringly similar, and rooted in Egypt. An Egyptian text written in 700 BC records a version of the creation myth from the time of the foundation of Memphis (the Old Kingdom of Egypt). It is strangely familiar: Ptah's heart thought of each thing as he created it, and his tongue repeated what had been thought of. His heart is the spirit, his tongue the word. His body is the universe. The myth contains within it roots of Christianity and Judaism and Islam, of microcosm and macrocosm, and of the power of the word in magic and in reality: 'In the beginning was the word, and the word was with God, and the word was God.'

Everything that the non-physical heart is to us now, it was to the Ancient Egyptians then. Here is a story, one of the oldest in existence, written down about 3,500 years ago but probably much older than that. It introduces heart ideas that echo through cultures and down the centuries.

The Tale of the Two Brothers

Anubis and Bata were brothers. Anubis was the elder; Bata worked for him and was like a son to him. All was well until one day Anubis's wife looked at Bata and 'her heart recognised him, as you recognise a young man'. She invited him to lie with her, whereupon he became 'like a cheetah of the south in his hot rage' and rejected her. Angry, she rubbed herself with grease and ash so she looked bruised, and told her husband that Bata had attacked her. Anubis sharpened his knife and hid behind the door of the stable to kill his brother when he returned with the cattle. The cows saw Anubis there and as Bata approached, they warned him. Terrified, Bata fled, calling on Pre-Harakhti to protect him. The god laid down a body of crocodile-infested water between Bata and Anubis, and from the far side Bata cried out to his brother that he was going to the Vale of the Pine – to Osiris's land across the water – to die, in effect.

The following morning Bata told Anubis his side of the story, and then 'he took a billhook for cutting reeds, he severed his phallus, he cast it into the water, where the electric catfish devoured it'. At this, the elder brother cursed his heart, and Bata said: 'I shall take out my heart by magic and place it on the top of the flower of the pine; and when the pine is cut down and my heart falls to the ground you will come to seek for it.' Bata told Anubis to put the heart in fresh water when he found it, and he, Bata, would return to life. In the meantime, so long as the heart was safe atop the flower, Bata was safe too.

Meanwhile the gods gave Bata a wife, a woman so beautiful that even the Nile desired her. When she washed in the Nile her sweet scent went down river and got into the Pharaoh's laundry. He searched her out and took her as his principal favourite. She told him about the pine and the heart, and made him have the tree cut down to destroy Bata.

When this happened, Anubis's beer frothed and his wine became clouded; thus he knew it was time to search for his brother's heart. After many years, 'he found a fruit, and lo! It was the heart of his younger brother'. He brought a cup of fresh water and placed the heart in it. By nightfall, the heart had absorbed the water, and Bata, lying dead on his bed, trembled in all his members, and he gazed fixedly at his elder brother, while his heart was in the cup. 'Anubis seized the cup of fresh water in which was the heart of his younger brother, who drank and his heart was in place and he became as he was before.'

Bata became a bull and went to court to frighten his wife; when she saw him, she had his throat cut. Where drops of blood fell to the ground, two great and beautiful persea trees sprang up. When she sat beneath one, Bata murmured to her, so she had the trees cut down and made into coffers. As the carpenters cut the wood, a splinter leapt into her mouth, and 'she perceived that she had conceived'. She bore a son, to whom, when he came of age and inherited the crowns of Egypt, Bata told the whole story. The wicked wife was punished; Bata lived happily ever after.

This story seems strange to modern eyes, but it brings up many, many aspects of the heart that recur throughout history. It speaks of eating the heart and washing it in water in order to win life after death, prefiguring the Christian sacraments of baptism and holy communion. It also foreshadows the French troubadour school of heart-eating poems, where wives are fed the heart of their lover by their jealous husband, as well as the heart-shaped chocolate boxes of today. The correlation between the heart and the phallus – the love/sex connection – is clear. There are links with the legend of Osiris, who was killed by his brother and dismembered before being brought back to life by his sister/wife Isis; and with Christ's abuse (His heart is pierced by the spear of the solider Longinus), death and resurrection; and with Dionyus, who was dismembered by Titans and rescued by his father Zeus, who snatched his heart and brought him back to life by feeding the heart in a glass of water to the nymph Semele.

The tree starts a long association of the heart with plants, where, for example, Christ portrays himself as a vine, or a fruit on the tree of the Cross, and is shown squeezing wine/blood from the grapes that grow from his heart wound. The 'flower of the pine' is the first link between the heart and the flower that developed into the Virgin Mary's heart, becoming the rose without thorns. Christ's heart is the rose with thorns, the same ones that crowned him at the crucifixion, representing the agony and the ecstasy of his passion, the promise of rebirth, and later making the fragrant and painful connection with romantic love. Petals represent blood, as in ancient legend. In Sufi poetry as well, the rose and the heart have a close and mystical connection; in Buddhism the heart is a lotus. Trees or human souls sprouting from fallen drops of blood, the soul separating from the body, either to spy or to be kept safe, and the heart as a fruit, appear in all folklores (there's a very funny West African one about monkeys and mangoes). The story foreshadows the virgin's conception, the treacherous kiss, the sacrificial death of a son, and his resurrection. Already, 5,000 and more years ago, the heart embodies identity, life, fertility, loyalty, God and love.

It is perhaps not surprising that the emotional territory of the Ancient Egyptian heart was as wide as, or even wider than, our own: 'Man's heart is his life-prosperity-health!' It could be happy or frivolous, 'big', meaning proud, or 'great', meaning high-minded and magnanimous, or 'wide', meaning patient. You could be master or mistress of it; it could be inflamed – a 'hot-heart' is a person out of control – or it could be content. The heart could ponder the truth, or even rest on Maat, the feather of truth. It could prevent its owner from hearing words of wisdom. When you studied, you put what you had learned in your heart (you learned it 'by heart'). Queen Hatshepsut built 'two obelisks of electrum whose summits would reach the heavens' in honour of her father Thutmose I and her heavenly father Amun. She entered, with a loving heart, into the plans of Amun's heart, and then when the obelisks were completed, her heart turned 'to and fro, in thinking what will the people say'; her heart encompassed love, belonging, consideration and pride. The love was to do with duty and devotion as much as romance.

The heart could love ('Be steady when you think of him, my heart, do not flutter!'). It could desire, fear, rejoice, worry and be sad. Because it could be filled with these feel-ings, it came to be seen as a receptacle: a jar, or a box. It could be appeased by kind words; it could incline to accept them; it could be brave and bronze; it could be broken and weak; it could be lifted up and reaffirmed; it could be disobedient and stubborn; it could jump and be like a bird in its flightiness (my mobile phone uses an icon of a heart with wings). It could bathe, fill, empty, retreat. It could be greedy, troubled, angry; it could lean or turn towards somebody in sympathy, but it was dangerous for it to move further than that. You could put it on someone, or towards them, in love and desire. You could put it behind them, which meant you worried about them. To wash the heart meant to relieve your feelings, which is part of the process of being reborn, and not a million miles from baptism. Intelligence, understanding, knowledge, memory and various gods lived there. It was where memories were kept (or lost). Above all, a heart could be delighted: Harkhuf, a sixth-dynasty official buried at Aswan, recorded in his tomb at length how the heart of King Neferkare Pepi II was gladdened by the prospect of seeing the dancing pygmy that Harkhuf had brought from the land of the horizon dwellers.

3 Juan Correa (Mexico, 1645–1716), *Allegory of the Sacrament*, 1690, oil on canvas, 164 × 106 cm

The vital importance of the heart comes up over and over again in the sayings of Ptolemaic instructors:

> The impious man who is proud of himself is harmed by his own heart.
> Do not let your tongue differ from your heart in counsel when you are asked.
> The heart cannot rise up when there is affliction in it.
> Wine, women and food give gladness to the heart – he who uses them without loud shouting is not reproached in the street.
> Do not open your heart to your wife or your servant; open it to your mother, she is a woman of discretion.

For the heart then, as now, contained our secrets and mysteries:

> One does not discover the heart of a man if one has not sent him on a mission
> One does not discover the heart of a friend if one has not consulted him in anxiety
> One does not discover the heart of a brother if one has not begged from him in want . . .
> One does not ever discover the heart of a woman any more than one knows the sky.
>
> From the *Insinger Papyrus*

Then, as now, romantic love was in the heart. And then, as now, love was foolish:

> My heart thought of my love of you
> When half my hair was braided;
> I came at a run to find you,
> And neglected my hairdo.
> Now if you let me braid my hair,
> I shall be ready in a moment.
>
> From *Papyrus Harris 500*

When an Ancient Egyptian died, the heart that had kept him physically alive came into its own. In the mummified corpse, the heart remained inside the body. As holder of a person's individuality, it had a vital role to play in his or her continued existence after death. As soon as someone died, they proceeded to the Hall of Two Truths. Here, before Osiris and a panel of other gods, the heart was weighed in scales against the feather of truth, emblem of the goddess Maat. If your heart was not there to be weighed, there was no way for you to be judged worthy of eternal life, and you could not get to heaven. Having your heart in the right place at that moment was very, very important.

The complex and poetic process of judgement was governed by the *Book of the Dead*, an ever-developing collection of spells and prayers, morality, magic and religion,

4 Judgement scene (detail showing the weighing of the heart), in *Book of the Dead of Kerqun*, Ptomelaic, Egypt, papyrus, height of roll 39 cm

intended to ensure eternal life. By the time of the twenty-sixth dynasty, you could buy a scroll of it and have your name inserted. According to this book, forty-two specific sins had to be denied, each to a specific god or demon:

> O Wide-of-stride who comes from On, I have not done evil.
> O Shadow-eater who comes from the cave, I have not stolen.
> O Lion twins who come from heaven, I have not trimmed the measure.
> O Cave-dweller who comes from the west, I have not sulked . . .
> O High-of-head who comes from the cave, I have not wanted more than I had.

And further sins have to be denied in general:

> I have not mistreated cattle.
> I have not sinned in the Place of Truth.
> I have not known what should not be known . . .
> I have not deleted the loaves of the gods.
> I have not eaten the cakes of the dead.
> I have not taken milk from the mouth of children.
> I have not held back water in its season.
> I have not dammed a flowing stream.
> I have not quenched a needed fire.
> I am pure, I am pure, I am pure, I am pure.

Whether or not you were deemed pure was up to your heart. As it was weighed – the jackal-headed god Anubis did the weighing, and Thoth, the ibis-headed inventor of writing and history, recorded the result – the heart told the gods whether or not you had been good. To persuade it to offer a good report of you, a scarab made of green stone, basalt or haematite would be laid inside the heart, or on the chest above it. Chapter XXXB from the *Book of the Dead*, pleading with the heart, would be either inscribed on the scarab or recited over it:

> O my heart [*ib*] of my mother,
> O my heart [*haty*] of my mother,
> O my heart of my being,
> Do not rise up against me as witness,
> Do not oppose me in the tribunal,
> Before the great god, the lord of the west [Osiris] . . .
> Lo, your uprightness brings vindication!

Your heart, here representing your conscience, could speak against you, while you – the rest of you – appealed to it for mercy as if it were a separate individual: it could save you or condemn you to death. (The heart was often referred to in the third person, as a friend; you could take counsel with it.) This separation is another reason for the constant declaration in funeral texts that 'My heart is mine.' Sometimes they read: 'His heart is his, his heart does not say what he has done' – which slightly gives the game away. If I were Osiris, I should want to take that heart to one side and have a word with it. But the heart was always portrayed as faithful, and the scales evenly balanced, in the hope of inspiring a favourable judgement. Thoth answered: 'Behold, I hold to be right the name of Osiris; and his heart, also, hath appeared upon the scales, and it hath not been found to be evil.' All dead male Egyptians were referred to as Osiris, to associate them with the god's resurrection; women were called Isis or Hathor. This is a simple imitation of God that continued in most religions. The entire ritual of mummification was an imitation of the death and resurrection of Osiris, that the dead might be given back their powers and abilities in the afterlife.

After the heart-weighing, Horus leads the deceased to Osiris and, though we are never shown Osiris's response, it is generally assumed that the virtuous dead will then proceed to his domains and live happily for eternity as a god. If you are found guilty, on the other hand, Ammit, the Eater of the Dead, with 'the forepart of a crocodile, the hindquarters of a hippopotamus, and the middle part of a lion', will eat your heart there and then.

The weighing of the heart is just one image that reappears in the Christian and Islamic ideas of Judgement Day. Take the thirteenth-century tympanum of the cathedral at Bourges, for example: a beautiful angel, St Michael, holds a pair of scales, weighing a chalice containing the good actions of the happy-looking child whose head the angel is caressing. On the other side, a nasty little bat-eared demon is desperately trying to trim the measure (one of the sins that had to be denied by the heart in the *Book of the Dead*). Standing around are various bigger demons, busy pitching damned souls into the great cauldron, bubbling on the flames of hellfire. The image of the scales was used by St Augustine, the Church Fathers and St John Chrysostom. It first appeared in the Middle East and came from there to Europe, and on into everyday modern usage: light-heartedness and heavy-heartedness may no longer be quite so closely associated with sin and guilt, but the image persists.

5 *Last Judgement* (detail showing St Michael weighing souls), c 1270, tympanum, front central portal, Bourges Cathedral

After your heart has vouched for your virtue, it is washed with cool water and Anubis gives it back to you (remember Bata?). Water is a straightforward image: it is thirst-quenching, life-giving, refreshing, cleansing. It is also an ancient symbol of returning to the womb or to the primordial waters out of which, in almost every culture, the earth itself was born. In Egypt it echoed the flooding of the Nile, which revitalised the country. Jeremiah, the most cardiocentric book of the Old Testament, also reflects the importance of washing: 'O Jerusalem, wash thine heart from wickedness, that thou mayest be saved'; as do the Psalms: 'Wash me thoroughly from my iniquitie, and clense me from my contrite heart . . . create in me a clean heart.' Modern Christianity's tidy little version of the plunge into water is water on the forehead in baptism; it correlates with the water that flowed from Christ's wound at the crucifixion.

6 Anon, a landscape with the Sacred Heart floating in a lake, c 1800?, coloured cut paper with watercolour, 56 × 82 cm

Weighed and washed, the heart waits until Isis calls the dead back to life. For the dead to be reanimated, it is vital that their hearts are not lost or hurt. They will need them so that they can remember their lives. This is another reason for having a heart scarab: as a spare. This comes from Chapter XXVI of the *Book of the Dead* (from the *Papyrus of Ani*):

> May my heart [*ib*] be with me in the house of hearts!
> May my heart be with me in the house of hearts!
> May my heart be with me, and may it rest there, or I shall not eat of the cakes of Osiris on the eastern side of the Lake of Flowers, neither shall I have a boat wherein to go down the Nile, nor another wherein to go up, nor shall I be able to sail down the Nile with thee.

The earliest Chinese medical texts, 3,000 years old, also knew that the pulse was significant, and had a complex and poetic set of medical theories. The heart controlled the *shên*, the spirit or 'divinely inspired part', and played the role of the king or master – *xin zhu* – in charge of the blood. Again, the spiritual and the physical were closely

7 Heart scarab, Third
Intermediate Period, Egypt,
serpentine, length 6.97 cm

bound. Ailments were attributed to disharmony and lack of balance, caused by the wind and the weather, 'noxious emanations' in heaven, or by wrong living.

> The heavenly climate circulates within the lungs; the climate of the earth circulates within the throat, the wind circulates within the liver, thunder penetrates the heart, the air of a ravine penetrates the stomach, the rain penetrates the kidneys . . . violent behaviour and scorching air resemble thunder.

In India, early knowledge was mostly magical, poetical and optimistic. According to the *Upanishad of the Embryo*, a Hindu scripture of around the eighth century BC:

> The establishment of the seven *dhatu* [blood, flesh, fat, tendons, bones, marrow, sperm] occurs in the heart. In the heart there is an inner fire, and where there is fire, there is bile, and where there is bile, there is wind, and where there is wind there goes the heart. The inner fire gathers in the form of bile, and is fanned by the wind (which blows through the respiratory and digestive systems). The heart is the seat of breath, and of blood vessels.

The *Atharvaveda*, a collection of 731 hymns in twenty books, incorporated herbal drug remedies, goblins, evil spirits, sorcery, archaic medical history, popular practices of the time and the first suggestion of circulation in the vessels. Here is a chant to get rid of jaundice and *hrddyota*, a 'heart affliction' that may or may not have been angina pectoris. The patient is to drink water mixed with hair from a red bull, and to wear a piece of the bull's skin, soaked in cow's milk and anointed with ghee, as an amulet round his neck.

Up to the sun shall go thy heart-ache and thy jaundice; in the colour of the red bull do we envelop thee! We envelop thee in red tints, unto long life. May this person go unscathed, and be free of yellow colour! The cows whose divinity is *rohini*; they who, moreover, are red – in their every form and every strength we do envelop thee. Into the parrots, into the thrush do we put thy jaundice, and furthermore into the yellow wagtail do we put thy jaundice!

The spiritual role of the heart was just as significant as in the West, and has many similarities. The heart *chakra* in Hinduism and Buddhism is the central sphere of spiritual power in the body, where the higher and lower meet. It unites man and god, and is the home of emotions. In the *Upanishads*, 'Supreme heaven shines in the lotus of the heart', and that is where you go to meditate. Various gods live there. As in Europe, the literality of this image caused beautiful, poetic confusions in anatomical beliefs.

It was the Ancient Greeks who divided the body and the soul, and started the process of sorting out the relationship between them. They were not immune to the primeval reasons for valuing the heart: its warmth, centrality in the body, mobility and system of connections. To those reasons they added their own, based, for the first time, on physical observation and natural philosophy. Once this admirable Hellenic clarity was in place, the history of the knowledge of the physical heart might have developed into a simple narrative of ever-increasing knowledge and understanding, as scholars and physicians learnt from each other, but this was not to be: the heart itself may seem relatively simple, but it is the centre of a complex system and coming to understand that system is taking thousands of years.

For a start, the physical heart was no more separate from the spiritual for the Greeks than it was for any other culture. Its prime duty was to be the home of the soul. One important observation, made very early, was that you could be unconscious without being dead, and that this was particularly likely to come about if you were hit on the head. The Greek system dealt with this by dividing the soul into two: the *psyche*, which was your individuality and personal immortality, and lived in the head, and the

8 Kalighat artist, near Calcutta, Hanuman, the Hindu monkey god, revealing Rama and Sita in his heart, nineteenth century?, India, watercolour with pencil and silver on paper, 45.8 × 28 cm

thymos, which was your heat, motion and physical life, your 'breath soul', and lived in the heart. At death, your *thymos* would return to the *pneuma* – the life spirit of the world, mankind and the universe – and your *psyche* would live on. This system accepted that even if your *psyche* were damaged, your *thymos* could continue to keep your body alive: 'The lights are on but nobody's home.'

The Greek philosopher Plato (c 427–347 BC) was most interested in the heart as one of the seats of the soul and as the centre of his philosophical microcosm, but he also reported a selection of traditionally held beliefs about the body: that blood was made in the liver and distributed to the body, which made flesh of it and otherwise used it up, and that the soul – *pneuma* – was breathed in from the air, and carried in the vessels. He also introduced the idea – which lasted in Europe until the end of the sixteenth century – that the body existed primarily as a vessel for the soul, and could not be considered separately from it (see Chapter 5). This was not a peaceful arrangement, particularly when it was adopted by Christianity: the body tormented and imprisoned the soul with its base desires; the soul was a bad guest in the body with its moral bossiness and yearning for release from the evils of the flesh.

The early Greeks assumed that the *pneuma* was drawn into the brain via the nose, and sent around the body in the vessels, possibly mixed with blood, and that a symmetrical, two-vessel system arose in the brain and went down the body like a tree (again recalling Bata and Anubis), stopping off at the heart, the liver and the spleen. Around 430–330 BC, a collection of medical works, known as the *Hippocratic Corpus*, came together. These works were associated with the Ancient Greek physician Hippocrates (c 460–370 BC), and the Hippocratic system of humours (see Chapter 5) pertained until the mid-nineteenth century, when the anthropologist Rudolf Virchow (1821–1902) developed the theory of cellular pathology, the basis of modern medicine. Like Eastern medicines, it held that health was balance and illness was imbalance; health was stability, illness was an upset. What had to be in balance were the humours or *chymoi*, the four fluids in the body. If too much of the blood (a plethora) flowing out (in a defluxion) from the heart gathers at the extremities, bearing an excess of one humour or the other, the subject will be made ill. Vomiting, diarrhoea, catarrh, pus and nosebleeds were seen as the humours trying to level themselves out, which suggested the remedy of bloodletting: tying a ligature around a limb, allowing the vessel to swell up and then lancing it to relieve the pressure, the excess humour and therefore the sickness.

The Hippocratics declared quite clearly that the heart was a muscle with two ventricles and two auricles wrapped in the pericardium. The right ventricle fed blood to the lungs and received air in exchange. The left ventricle, containing only air, was the seat of the innate heat, which generated the humours from food, kept them in balance and moved them around the body. The vessels 'are the springs of man's existence, from them spread throughout his body those rivers with which his mortal habitation is irrigated, those rivers which bring life to man as well, for if ever they dry up, then man dies'. Inside the heart are:

> the hidden membranes . . . a piece of craftsmanship deserving description above all others. There are membranes in the cavities, and fibres as well, spread out like cobwebs through the chambers of the heart and surrounding the orifices on all sides and emplanting filaments into the solid wall of the heart . . . these serve as the guy-ropes and stays of the heart and its vessels, and as foundation to the arteries. Now there is a pair of these arteries, and on the entrance of each three membranes have been contrived, with their edges rounded to the approximate extent of a semi-circle. When they come together it is wonderful to see how precisely they close off the entrance to the arteries.

Clearly, someone had seen inside a heart, and seen the valves. They had noticed that veins and arteries are different: arteries (from *aer tereo* – carriers of air, as they were believed to be) are much tougher. Plato's student, the philosopher Aristotle (384–322 BC), who was the first consciously and decisively to base systematic anatomical theories on dissection (albeit animal, not human), distinguished between their structures but not their functions, and for him, both – and indeed everything else – originated in the heart. The Hippocratic view was that the heart was the seat of man's intelligence, 'the principle which rules over the rest of the soul', but they were quite firm: 'that the source of our pleasure, merriment, laughter and amusement, as of our grief, pain, anxiety and tears, is none other than the brain'. Aristotle did not agree. The heart was the prime mover of life, he said, and 'the motions of pain and pleasure, and generally of all sensations, have plainly their source in the heart, and find in it their ultimate termination'. All the blood vessels originated there, and blood was made there.

Dissection was of great help to the advance of anatomical knowledge, but it was not without flaws. The dead, immobile heart looks nothing like the flexing living organ. Also, the method of killing an animal affects the nature and amount of blood in its heart and vessels. Strangling (which Aristotle favoured) stops blood in its tracks; throat-cutting drains at least some off. This is why arteries seemed to contain not blood but air. No one could know that this was not the case when the creature was alive – it seemed more likely to the cerebrocentrists that the tough, seemingly empty vessels contained invisible *pneuma*. Moreover the foetal heart – which is probably the only kind of human heart Aristotle ever saw – is different from the adult heart. Animal hearts can also be very different from human hearts. And even when you know how many chambers there are, it can be hard to make them out: cutting through different planes can produce very different-looking cross-sections.

The first known and recorded human dissections took place in Alexandria in the third century BC. Alexandria was a cosmopolitan city, where Egyptian traditions of mummification and Greek philosophical theories (which held that only the soul, not the body, could expect immortality) combined with the presence of a medical school in a society where slaves and condemned criminals were not considered entirely human – except in body. As a result of the opportunity this threw up, the Alexandrians learned to distinguish between blood vessels and nerves; and they recognised the systole – contraction – to be the active phase of the heart's action, by which blood is expressed from the heart. They also discovered the central nervous system, and recognised through experiment that it was the organisational centre of the body. So much for Aristotle's central and sensory heart. Using Aristotle's ideas of working rationally and through observation, the Alexandrians displaced his theories.

A constantly shifting balance between what was observed and what was assumed to be the case greatly increased the length of the voyage to understanding. Thus the anatomist Erasistratus (330?–250? BC), one of the great Alexandrians, could believe that the arteries contained only vital spirit, even though they bleed when you cut them, rather than ooze spirit. The finest veins, he surmised, too fine to hold blood, intercommunicated with the finest arteries, which when damaged lost their *pneuma*, and drew blood from the veins into the resulting vacuum. This is only one splendid example of cutting your anatomy to fit your theory. And yet, the finest arteries do link up with the finest veins, via the capillaries, and how this works was not solved until after Harvey discovered circulation, and then only with the aid of a microscope.

Erasistratus conducted another interesting experiment: when Prince Antiochus of Syria was melancholy, and his pulse slow, the doctor had the ladies of the court parade past his couch. When the prince's stepmother Stratonice appeared, his pulse picked up, and when dispensation was given for them to marry, sphygmic stability was achieved. We know all sorts of things now about hormones and blood chemicals, adrenalin and serotonin, but the pounding of our hearts has always spoken to us directly about love and passion.

The Greek physician Claudius Galen (c 129–200 AD) used merely to recite the names of likely sweethearts while holding the wrist of patients he suspected of being lovelorn (the name Pylades produced a 'turbulent pulse' in one heartsick wench). Galen, known as the Prince of Physicians, was the son of an architect and a woman who, he said, used to bite her serving maids. He studied in Alexandria, but the physicians had been expelled by the Ptolemy of around 200 BC, and for medical students it was back to animals (apes, pigs, goats and an elephant), skeletons and fortuitous circumstance. Galen revised the old image of the tree by upending it: he rooted it in the liver and had it grow upwards towards heaven. He also held, with the Alexandrians, that the brain, not the heart, was the source of the nerves, and of their powers.

In his book *On the Causes of Pulsation*, Galen described the body as he saw it. He held that there were in effect two separate blood systems: nutritive (venous) based on the liver, and respiratory (arterial) based on the air-breathing heart, which goes back to the traditional view of the two main vessels running the length of the body. The heart was to him a single organ divided in two to deal with the two systems. He proved by experiment that arteries contained blood, not just pure spirit, and postulated tiny holes in the septum between the ventricles, through which a little venous blood could seep, in order to undergo the vital process of mixing with *pneuma* to make vital spirits, which would carry heat through the arteries to the body. Galen's ideas dominated knowledge of the heart for centuries. They were satisfying, and nothing new was discovered – indeed much of what had been discovered was forgotten after the fall of the Roman Empire. Dissection was no longer, Greek texts were neglected. The Dark Ages descended on Europe.

What we think of as Greek culture, which in the West remained lost for generations, was preserved and naturalised in its Arab branch by Arab scholars, who were of necessity interpreters, compilers and reconsiderers of the theories of Hippocratic, Galenic, humoural medicine. 'The veto of the religious law and the sentiments of charity innate in ourselves alike prevent us from practising dissection', as the physician Ibn al-Nafis (1213–88) put it. Despite this limitation, Ibn al-Nafis was the first to say that Galen was wrong about his septal pores, 'for the substance of the heart is solid, and there exists neither a visible passage, as some would suppose, nor an invisible passage which will permit the flow of blood, as Galen believed'. He agreed with the rest of Galen's physiology, and came up with a solution: blood crossed the lungs 'so that it can spread out in their substance and mix with the air' and thence to 'the venous artery, and from there to the left ventricle'. This came to be called the pulmonary transit. It is exactly what happens and Ibn al-Nafis was the first to identify it.

The Persian physician and philosopher Ibn Sina (978–1036), known in the West as Avicenna, meanwhile suggested that the all-important heart might act through other organs – the brain for psychic functions and the liver for nutrition. This was the first glimmer of a way of reconciling Aristotle's views of the heart with Galen's. In a micro/macrocosmic way, it also makes room for the idea that God, rather than being omnipotent, acts through humans; for individual responsibility over fatalism.

Only when the idea emerged that scientific knowledge could be cumulative did scholars give precedence to the methods, rather than the findings, of Galen and Aristotle, and start again to use reason and observation, and to make progress. Medieval scholars tended to feel inferior to the great ancients, not only intellectually – which made them doubt their own understanding rather than question the Greeks' authority – but also, as dissection again became possible, physically. In 1302, post mortems were allowed for legal and medical purposes in Bologna; by the 1400s, the universities there, as well as in Padua and in Florence, were legally allowed to dissect. But when the human body on the slab contradicted Galen, the German anatomist Franciscus Silvius (1614–72) declared that the body was wrong.

So yes, it was about time for the Renaissance to begin. In a long spree of shared, published and debated learning between anatomists, Hieronymus Fabricius (1537–1619) identified and analysed the valves, Andreas Vesalius (1514–64) doubted Galen's septal pores, Realdo Colombo (c 1516–59) denied them, Michael Servetus (1511–53) conceived of the pulmonary transit whereby blood goes to the lungs and back for oxygenation, and Colombo proved it. Galen was wrong, and this was a great shock. Everything was in place for Harvey to discover the circulation of the blood. 'Damme, there are no pores and it is not possible to show such', he declared, and in a grand flourish, he reversed the flow of blood in the veins so that it returned to the heart, and the rest fell into place. Not, of course, that everybody accepted it easily. It was a *great* shock.

The spiritual and metaphorical roles of the heart survived all this unblemished. Indeed, they took on aspects of physicality. Plato's use of the term 'chamber', for

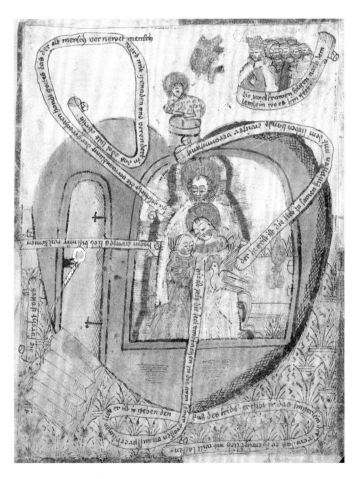

9 Anonymous nun, the heart as a house, fifteenth century, German, pen drawing

O beatam cordis ædem!   Animose puer verre,
Te cui cælum dedit sedem   Monstra tuo vultu terre,
Purgat suis manibus.   Tere tuis pedibus.

Anton. Wierx fecit et excud:

10 Anton Wierix (Belgium, 1552–1624), Christ clearing demons out of the believer's heart with a broom, c 1600, engraving with etching on paper, 7.8 × 5.6 cm

example, combined with the ancient tradition of gods living in the heart of man to produce the charming image of the heart as a house, which could be looked after and kept clean of sin and wickedness. 'The house of my soul … is in ruins', wrote St Augustine; 'restore it'. 'There are reception rooms and bedchambers in it, doors and porches, and many offices and passages', wrote St Makarios in the fifth century. In Islam, as the Sufi Muhammad al-Ghazzali (1058–1111) tells us, 'the heart has a window which opens on the unseen world of spirits'. A fifteenth-century nun whose name does not survive made a series of enchanting pictures of Christ's heart as a house with her sitting inside it, on his knee: the aorta is the chimney, there's a staircase or a ladder of virtues leading to the door, and the lamb of God is sitting on the roof.

The window and door are very significant – they are versions of the wound, through which the love of God can come and go. When Christ's heart was pierced on the cross by Longinus's spear, it not only released blood and water – represented in the two sacraments of holy communion and baptism, themselves representing God's love for man – but it also echoed the piercing of the lover's heart by the wicked little Roman god of love Cupid, with his bow and arrow, which releases carnal love. This is why the tattoo heart with the arrow through it can say either 'I Love Mum' or 'Bob 4 Lou' without being inappropriate. The suffering of the heart was generally held to be a good thing – an imitation of Christ, and an opening up.

And why is the heart written on anyway? Because in the book of Exodus, when God gave the law to Moses – 'two tables of testimonie, tables of stone, written with the finger of God' – he told him that the words of the law were to be kept in the hearts of men. The two tablets of stone lent themselves easily to the two sides of the heart, and from this arose a strong and lasting image: the written-on heart. From this, and from the ancient beliefs that memory and knowledge lived in the heart (sometimes kept on rolled-up scrolls), evolved the complex and charming idea of the heart as a book. Writing on actual hearts followed swiftly on: saints had Christ's name written on their hearts; lovers that of their sweetheart. Mary Tudor had hers inscribed 'Calais'; Robert Browning, 'Italy'. Poets from George Herbert to Ted Hughes have used the image. And it is a painful image, a form of wounding, and often a phallic one: the quill pen is related not just to Cupid's arrow and Longinus's spear, but to any other long hard thing that pierces something soft and red and releases love while laying claim to it. Just look at St Teresa.

This writing and piercing has a sacrificial element. The heart is a perfect sacrificial object: so rich, so symbolic, so compact. The Aztecs of Mexico, as every schoolboy knows, sacrificed human beings, tearing out their hearts on pyramids so that the sun, on which all life depended, could be nourished with the heart's blood. More precisely, they were recycling a vital energy, *tonalli*, 'the warmth of the sun', which exists in the universe and in all living things, and which is squandered by human beings when they die, and wasted on their difficult journey to Mictlan, the place of the dead. Regulating death – sacrifice – harnessed the *tonalli* and allowed it to be used for the benefit of the living instead.

There were eighteen months in the Aztec year; each had a feast, lasting its whole length. The rituals were astoundingly complex. Fray Bernardo de Sahagun (1499–1590), a Franciscan who recorded twelve volumes on Aztec culture from 1529, tells us that for the festival of Tezcatlipoca, for example, a man was chosen to impersonate the god for a year before being sacrificed, and he lived a life appropriate to a god. The list of his perfections goes on for several pages, starting with 'of fair countenance, of good under-standing, quick, of clean body, slender, reedlike, like a stout cane, like a stone column

11 Giovanni Lorenzo Bernini (Italy, 1598–1680), *Ecstasy of St Teresa as the Angel Prepares to Pierce her Heart* (detail), seventeenth century, marble

12 Joseph Florimond Loubat (USA, 1837–1921), *Magliabecchiano Codex*, XIII.3, Danesi, Rome, 1904, facsimile of post-Colombian Mexican manuscript in Biblioteca nazionale, Florence

all over' and going on, 'smoothed, like a tomato, with straight hair, no pimples, with a forehead not shaped like a tomato or a bag, not swollen-cheeked, bulging-eyed, bent-nosed', and so on. 'He was importuned, he was sighed for, there was bowing before him.' He went about playing the flute, with eight servitors. He wore a stole of popcorn flowers, golden shell earrings, earplugs of turquoise, a seashell breast ornament, gold and turquoise bracelets all up his arms, golden bells on his legs, and obsidian sandals with ocelot-skin ears: 'Thus was arrayed he who died after one year.' When he was killed, his heart was taken and raised up in dedication to the sun. 'And this betokened our life on earth. For he who rejoiced, who possessed riches, who sought, who esteemed, our lord's sweetness . . . thus ended in great misery . . . No one on earth went exhausting happiness, riches, wealth.'

Here is a description of a sacrifice to the rain gods:

When they slashed open the breasts of the victims . . . they seized his heart; they went to place it in a vessel painted blue, named the cloud vessel, which was painted with rubber on four sides, and its accoutrements were papers dotted with drops of liquid rubber, much rubber . . . All their hearts they continued casting there . . . And when they went to arrive in midwater . . . there they brought the boat in . . .

thereupon trumpets were played. The Fire Priest arose in the prow of the boat. Then they gave him the cloud vessel, which went filled there with hearts. Thereupon he cast it in the midst of the water. It immediately was swallowed, it immediately pierced [the water]. And then the water foamed, kept surging, roared, crackled continually, crackled as it surged. Bits of foam formed.

It is easy to syncretise the bloody sacrificial Aztec heart to the Catholic Sacred Heart of Jesus (see Chapter 4). As an idea, it arrived in the New World from Europe in the sixteenth century, though as an image probably not until 150 years later. All the same, when the Spanish monks talked of Christ's heart, and of men directing their hearts towards Christ, of hearts 'full for God', and of Christ's blood and body 'given up' and consumed, they were on familiar metaphorical ground. The Aztec Sun God drank the blood of his human sacrifices; Christians drink the blood of God, made human and sacrificed. The Aztec made effigies of their seed god out of amaranth seeds and sacrificial human blood, which they then ate; Christians eat the body of God in a piece of bread. If we compare the image of the heart in the lake on page 14 with the description of the sacrifice above, we can see that one of Christianity's

13 Offertory box, eighteenth century, Spanish, polychromed wood, 30 × 20 × 160 cm

greatest talents was incorporating religions that had come before it, or stood in its way. As early as 1688 the rites of Huitzilopochtli were being compared to the Christian sacrament by Sor Juana Inés de la Cruz (1648–95) in her play *The Divine Narcissus.*

Seen in this light, Christianity does look disconcertingly like a religion of human sacrifice. Sacrifice means 'made sacred'; the Aztec sacrificed man imitates – becomes – the god before dying. In Christianity, God becomes man before dying. In 567 the Council of Tours had to forbid the arrangement of the host in the shape of a human body, and in 1215 Pope Innocent III declared that transubstantiation was a fact: the wafer *was* the flesh and the wine *was* the blood (not represented: *became*). During the early years of the Inquisition, one of the horrendous crimes of which Jews were accused was torturing the consecrated wafer: the wafer was held to feel pain. A thirteenth-century mould that made communion wafers in the shape of a heart survives in Spain.

The heart, of course, has a very special place in Mexican art. For example, in her 1951 *Self Portrait with the Portrait of Dr Farill*, the heart of Frida Kahlo (1907–54) lies on her palette like a pile of red paint shot through with blue, and she holds a bunch of paintbrushes, sharp as darts or surgical implements, and dripping red – another form of the written heart. The ways in which ancient heart imagery reappears in the modern world never cease to delight me. The 'St Louis Blues' sings of the heart like a rock (Moses's tablets?) thrown into the sea. There's a country song called: 'My Tears Have Washed "I Love You" From the Blackboard of My Heart', which goes straight back to the written-on heart, and also reflects the early Christian concept of the Book of the Heart, where good and bad deeds and thoughts are recorded, and can be washed off by God as writing could be scoured off vellum by a scribe. Blondie's song 'Heart of Glass' recalls the Sufi image of the heart as a mirror that must be cleaned so that God can see his image clearly in us, or the biblical injunction on 'the thoughts and intents of the heart . . . all things are naked and open unto the eyes of him with whom we have to do' (Hebrews 4). There's a sadly neglected soul song called 'Getting Mighty Crowded', in which Betty Everett sings of packing up her memories and moving on out of her lover's heart. She's going to 'leave the neighbourhood', and 'find another heart where she can live all by herself'. Janis Joplin is a willing Aztec when she wails, 'Take it, take another little piece of my heart now, baby, you know you've got it if it makes you feel good.'

But back to the tattoo. We've seen why it's a heart, why it has an arrow piercing it, why it has writing on it and roses round it. But why is it that shape? For centuries, hearts were drawn from descriptions, rather than observed, and a common description was that the heart was pine-cone shaped, and twisted to the left. Pine-cone hearts can be seen all over the place, often held by the fat end, upside down. Giotto's early fourteenth-century *Caritas* (Divine Love) in the Capella dell'Arena at Padua holds up her pine-cone heart to God for all the world as if she had just taken it from the fruit bowl in her other hand. It is large, and a tiny God in the corner reaches down with two tiny hands to take it from her. The heart of Caritas often flamed like later sacred hearts; the flames

14 Hendrik Goltzius (Netherlands, 1558–1616), Christ holds up a glass heart filled with animals and cures a sick woman with the sacramental blood and water from the wound in his side, 1578, engraving, 23.7 × 18.3 cm

came from what would later become the pointy end. By 1360, a *Caritas* by Giovanni del Biondo shows her heart held by the pointy end, slightly scalloped and with the flames coming out of the aorta like a gas jet. How did this change come about?

The first hearts with indentations in the top appeared in the early fourteenth century in northern Italy. Pierre Vinken, a Dutch cardiologist who has written a wonderfully detailed book following the process, suggests that the first scalloped hearts appear on a necklace round the neck of Love's horse in Francesco da Barberino's *Documenti d'Amore* from the early 1300s, and in an anatomical book by de Vigevano published in 1347. An *Amor* by Giotto of 1323 shows the blindfold boy himself with a string of conquered hearts strung about his torso like a bandit's Sam Browne; you have to look closely but they appear to be indented. Barberino studied in Bologna, a centre of anatomical study, so he would have been acquainted with developments in anatomy – including erroneous ones. Vinken connects the dent with the old question of the number of chambers in the physical heart. Ancient discussions of a smaller, third chamber in the middle, between the two obvious ventricles, led, through confused translation, to phrases such as Galen's *duos ventriculos, dextrum et sinistrum, et in medio fovum*, reappearing in Middle English as 'two ventricles, right and left, and in the middle a ditch or pit'. According to another description, from Italian professor of medicine Mondino dei Luzzi (1275–1326), the third chamber lay 'in the thickness of the septum'; would this mean inside the heart, or seen from the outside? Once a *fovea* or 'concavity' is seen as a ditch, it is clearly open on one side, so it cannot be within the heart. Even if it is inside the heart, it is only natural to make a central indentation on the top of the heart to show that the chamber in the middle is smaller than those on either side. Scribes who copied and translated the texts, on the whole knew nothing of anatomy, and would have had a job trying to reconcile all the descriptions.

Moreover, the heart had always been considered to have two sides (hence the pores through the septum to unite the two systems). This was quietly reinforced by the biblical image of the two tablets of the law, written in the heart. The broken heart also has a role here: a heart split in two is a contrite heart, a heart ready to receive God's love, a compassionate heart. The heart's symbolic role needed to be visibly dual. The dent achieved this, and the point emphasised it, and thus a beautiful, useful, symmetrical, unmistakable, flexible, universal graphic symbol became available to humanity.

I learnt something new about the heart recently. Despite years of interest and research, in the time-honoured tradition of heart studies I had failed to notice something obvious. In diagrams and descriptions, the heart is said to have two sides, left and right. In fact, when we see it in place, lying snugly in the human breast, we note that it is aligned front and back, with the side known as the right to the front, and the left to the back. Thus can we still be misled by what we think we already know. And as Proverb 15 says: 'The heart of him that hath understanding seeks knowledge.'

INTERVIEW

Sir Terence English

Sir Terence English

Pioneering cardiothoracic surgeon, in conversation with Melissa Larner

MELISSA LARNER: When did you first become interested in heart surgery?

SIR TERENCE ENGLISH: Quite early on during my medical-student days. I'd been an engineer first, and I was attracted to cardiology and the heart because of its mechanical properties. There were flows and pressures and resistances, which appealed to me. So I applied for, and got, the job of houseman to the senior cardiologist at Guy's Hospital, where I was working, and at the end of my six months with him, I decided to pursue a career in cardiology. He said that he thought I'd be a better cardiologist if I worked with the surgeons for six months, so he arranged for me to work as a senior house officer with Sir Russell Brock, as he then was, later Lord Brock, in the Thoracic Unit at Guy's. I started there in 1963 and it was an amazing experience. Open-heart surgery had only been going since 1955, and there were still lots of problems, both with the heart-lung machine that was used to support the circulation while one was operating on the heart – the technology of which was pretty crude – and also, just a simple lack of knowledge about some of the more esoteric cardiac diagnoses. Within a very short period, I realised that I wasn't going to go back to cardiology and that I'd much rather be a cardiac surgeon. The Thoracic Unit was an extremely stimulating place at the time. Brock had been one of the great pioneers, having developed closed-cardiac surgery in Britain. And then Donald Ross, a South African like me, who had just become a consultant, was very innovative and a superb technical surgeon, so they made a great combination. Having decided to be a cardiac surgeon, I first had to do general surgery, and then progress to specialist training. One of the great appeals was that in order to make a diagnosis, you used your physical senses of observation, palpation and listening. There was a very strong emphasis on clinical cardiology at

Ben Edwards, *Red* (blood reservoir on perfusion machine), 1996–9, C-type print, 152 × 122 cm

Guy's, and so I was trained in this, and it was fascinating to try and make a diagnosis on the basis of those skills. Then during the surgery, you'd find whether your diagnosis was correct or not in a very immediate way. So it was a great learning opportunity.

ML: Did it feel like a vocation?

TE: Yes, very much. I hadn't thought of doing surgery until that time. My own inclinations had been more those of a physician, but that's when I determined quite clearly that I was going to be a surgeon.

ML: Can you say something about attitudes to heart surgery in Britain when you were first working in this field?

TE: Yes. They were interesting. There was a lot of excitement among the cardiac surgical fraternity, who felt that they were on the edge of developing something very important, which would have a huge impact, because of the prevalence of heart disease. There was also a lot of conservatism among cardiologists, who had worked against the development of cardiac surgery for many years. Some believed that there was no role for surgery on the heart. There had been one mitral valvotomy performed successfully in 1923 at the London Hospital, but that surgeon never got another referral because the cardiologist at the time didn't believe that mitral stenosis was the cause of heart failure. And the early operations by Brock and those in America who were working on the heart at the same time were attended by very high mortality. To me, it seemed obvious that once one was able to operate within the heart, as a result of the development of the heart-lung machine, this was something that would have huge application.

ML: So how did it come about that heart surgery finally got its place?

TE: The two great cardiac surgeons who really put modern cardiac surgery on the map were John Kirklin at the Mayo Clinic in Rochester, Minnesota, and nearby, also in Minnesota, Walt Lillehei – a quite extraordinary man. Instead of using the heart-lung machine, Lillehei used the mother's circulation to support the circulation of the child while the heart was being operated on. So these two, in 1955, started programmes fifty miles apart, and the developments took over from there.

ML: And you did the first successful UK heart transplant?

TE: Yes. It was interesting because Donald Ross, who was a great mentor for me, did three transplants in 1968, soon after Christiaan Barnard's first operation. Barnard and Ross were contemporaries at Groote Schuur Hospital in Cape Town, and had trained together. I'd visited Barnard in 1963, just before starting on the Thoracic Unit, because Ross had suggested that as I was going to South Africa, I should spend a week with him and see what he was doing. I was not impressed with Barnard's technique, nor indeed his behaviour in the operating room. He was very excitable, and he'd shout and scream and say that nobody was helping him. But he was very meticulous and obsessional, and he was getting good results at that time. There's no doubt in my own mind that Ross was a far better surgeon. So Ross did three transplants, fairly soon after Barnard's: two at the National Heart Hospital, and one at Guy's. But all the patients died fairly early on, which was partly a reflection of the fact that very little was known about how to manage patients after transplants. After that, the Department of Health put a moratorium on further heart transplants in Britain. That was understandable, because if one looks back on the 160-odd transplants that were performed worldwide in the two years after Barnard's first operation, only 11 percent survived for two years. But Norman Shumway at Stanford University in California kept going, and Barnard was doing some operations in Cape Town. I visited Shumway in 1973, soon after I was appointed to Papworth Hospital in Cambridge, and I saw patients who'd received transplants and who were doing fairly well, and I thought that Britain ought now to have its own programme. So when I came back to Papworth as a fairly junior consultant, I discussed it with my senior colleague Ben Milstein, who was supportive, and then with Professor Roy Calne, who had a big liver transplant programme in Cambridge as well as doing a lot of kidney transplants. We decided to work towards establishing a joint programme, and thereby hangs a long tale, which has been described elsewhere. But eventually, in January of 1979, against the wishes of the Department of Health, who wouldn't fund or indeed recognise the work at that time, the local area health authority chairman, a wonderful woman, said that she'd be agreeable for me to use our facilities for two cases, and then after that we'd have to find our own money. The first operation, alas, was not successful. The patient only lived for seventeen days, which was a huge disappointment. He was very sick at the time, and he had a cardiac arrest before my assistant could put him on the heart-lung machine, while I was taking the donor heart out elsewhere, and he suffered brain damage. And so, although the transplant went fine, he had to be maintained on a ventilator and therefore died from infection. The second case, which was the first successful one in Britain, was in July of that year, and that was Keith Castle.

ML: Can you describe how that felt?

TE: Oh, full of emotion. With the first one, when we'd waited so long to get a donor heart, and the actual transplant had gone so well, it was just devastating to find that the patient wasn't going to survive because of the brain injury he'd suffered beforehand. But we continued to get support from the health authority, and so we did the second case. Keith Castle was a wonderful man. He was not an ideal patient from the medical point of view. In fact, he was pretty awful, because he was a heavy smoker and he had bad vascular disease in the legs and a peptic ulcer. But he had other things going for him, which was tremendous spirit and fortitude, and he did more to advance the idea and the acceptance of cardiac transplantation in those early years than anybody else. He was a builder – a cockney from Wandsworth – and with all the humour of a typical cockney. He just sailed through everything, and he lived for five and a half years, during which time he had a very good life. We were fairly lucky with the first six patients that I did: four of them were medium-to-long-term survivors. This was very important because there was a lot of opposition, both within the hospital – from two of the cardiologists – and from outside. The press was against it at the time, and I remember Bernard Levin wrote a very passionate article against cardiac transplantation.

ML: It sounds so crazy now, doesn't it?

TE: Exactly. The interesting thing was that I knew very soon that there had to be an important future for it. This was because it was so remarkable how you could have a patient who was in very severe heart failure, practically moribund, put a new heart into him, which allowed that heart to pump oxygenated blood around the body to all the tissues – the brain, the liver, the kidneys – and within as short a time as a week to ten days, you saw an enormous transformation in that individual. So although there were great disappointments and deaths – due to rejection, mainly, and infection – there were enough good survivors to keep our spirits up. Indeed, I heard yesterday of the death on Sunday of a patient whom I'd transplanted in September 1980. Gordon McDonald, therefore, had survived for just on twenty-six years, and I would never have believed at the time that this was possible. We were thinking in much shorter terms of five to ten years' survival. But he had a truly wonderful and active life, until the last two months, and he was able to see his children grow up, and his grandchildren.

ML: Do you feel that because of the emotional and iconic nature of the heart, you had a different relationship with your patients than, say, a renal surgeon?

TE: Yes, I think I probably did, but this wasn't just because of the heart and its central position within the body and its other connotations. Every tissue depends on a supply of oxygen and that's the heart's function: to provide the tissues with oxygenated blood, so it is absolutely central to the life of every being. But I think where the difference in relationship came in was that, because the cardiologists at Papworth weren't keen on becoming involved with this work, I actually assessed all the patients who were referred to me for transplantation by cardiologists up and down the country. I'd get them in for about five or six days. We had a very good social worker, who helped a great deal, but then I'd see them with their spouses – usually their wives because they were mostly men – and I'd go through their whole story right from the beginning. Often they had very long histories of heart failure and when they'd go to the clinic they'd see a different doctor each time. Their case notes would grow in size, so that whoever was seeing them in the clinic would say 'Oh, my goodness!', and it would be very daunting and wouldn't allow enough time for a thorough review. Whereas, because we knew our decision was so important, we'd look at their cardiac history thoroughly from the beginning, so they'd feel that we were absolutely committed to their welfare.

Once they'd been added to the transplant list, the next problem was to get a heart. I was absolutely frank with them about our statistics – most of the time, between two-thirds and three-quarters of the patients that we listed got transplants before they died. It was probably better than it is now. So they were told that, and they were warned that they might be brought in for false alarms and so on. We sent them back to their referring cardiologist, whom we asked to keep us informed of their state of health, so that if things became urgent we might give them a heart ahead of someone who had perhaps waited a little longer. There were many interesting ethical issues to be dealt with, and I was open with the patients about these. We also built up a very supportive team at Papworth, which made the patient feel identified, not just with me, but with the whole transplant team. This was a feature of the programme that I was proud of.

ML: So your special relationship was to do with the fact that they got attention and support from you, and not with how they felt about the heart?

TE: No, not entirely. There were some who'd say to me, 'Mr English, how am I going to feel if I get a woman's heart?', and I'd just have to reply, 'Well, the heart is a pump, it's not the seat of emotion, and as a pump, as long as it delivers the right amount of oxygen around your body and into your tissues, you'll feel fine.'

ML: So you don't give any credence to cellular memory, or those stories that patients tell about taking on the characteristics of their donors?

TE: No. One knew some had these ideas, but one tried to make them realise that this wasn't so. The other point that I think is worth mentioning is that with those patients who died before a heart became available we often got appreciative letters from the relatives saying that although they were very sad that their loved one didn't get a transplant, we'd given them hope at a time when they had no hope at all. Also, although donation of organs is always a very difficult, sensitive, traumatic thing for parents or spouses of donors to go through, there's no doubt that many of them got relief or satisfaction from knowing that out of such terrible tragedy some benefit would come to another person. That's one reason why I believe that donation should be voluntary.

ML: You said that it's even harder now to get donor hearts?

TE: The number of heart transplants each year has gone down dramatically over the last ten years or more, and this is purely because there haven't been enough donor organs. There are all sorts of reasons for this, but one factor is seat belts. Many of the early donors came from young people who were killed in road-traffic accidents. Someone with a severe head injury would be brought in, put on a ventilator and found to be brain dead. But if they were young and still had a good heart, the relatives would be asked to consider donation. Now, the incidence of mortality due to road-traffic incidents has gone down fairly dramatically. Other reasons I'm just not sure of. There have been significant attempts to try and increase the number of donor cards among the population. There have been innumerable programmes on television to try and increase organ donation, but the result has been disappointing. This is a worldwide phenomenon, although it's worth noting that Spain has the best rate of heart transplants per population in Europe, if not the whole world. They've concentrated on establishing a cohesive organisation of transplant co-ordinators in every major hospital, who are trained to be informed of any potential donor and then to seek permission from the relatives in a sympathetic way. They work closely with the intensive-care units because there's always extra work for intensive care if organ donations take place. So, the lack of donors has become quite a big problem and this has been a stimulus for the evolution of mechanical-assist devices.

ML: You did the first transplant using one of those, didn't you?

Yes, in Britain. I implanted a total artificial heart, a Jarvik heart, in 1986, as a bridge to a transplant. Robert Jarvik was an extraordinary man, who had developed an artificial heart. This had originally been the idea of a man called Willem Kolff, who had developed kidney dialysis in Holland during the War and then went to America. He developed a mechanical heart in Utah, which Jarvik perfected. Frank Wells, my colleague at Papworth, and I went to Utah to train in the use of this artificial heart, putting it into calves and then going back three months later, taking it out and putting a calf heart in. This was a difficult procedure. But eventually we were funded by Humana, an American organisation, to the tune of half a million dollars to intro-duce this into the NHS, and I did the first case in November 1986. The Jarvik heart, which ran off compressed air, was large and relatively clumsy, but it functioned well and the patient, who'd been very ill, began to recover. However, we were offered a human heart a few days later, and had to go through the laborious process of removing the artificial heart and inserting the human one. This was successful, but my experience with this case suggested that using a total artificial heart as a 'bridge' to transplantation was not a sensible option.

ML: Do you think artificial hearts are the future of heart transplantation?

TE: Recent experience with left-ventricular assist devices has been encouraging. These act as a support to the failing heart, which is left in situ. But total replacement of the heart with a mechanical device and a built-in power supply still seems to be a long way off. However, I think it's right that research in this area should be pursued by those countries that can afford it.

The other area that's being pursued is xenotransplantation – the use of organs from animals species. Here there has been some interesting research using geneti-cally engineered pig hearts for transplantation into baboons as a preliminary to their use in humans. This is an attractive long-term proposition, but there are still many hurdles to be overcome.

The Anatomised Heart
Understanding the Pump

Jonathan Miller

Human beings have always acknowledged that there is an association between existence and respiration. The start of life is traditionally identified with the first breath, and until more reliable signs of death were recognised, the misting of a mirror or the stirring of a feather were accepted as conclusive evidence that life still lingered. Most death-bed scenes include detailed accounts of the penultimate changes in respiration: the mourners seem to hang on the long, breathless silence, catching their own breath until the dying patient sighs once more.

Judging by the number of ways in which it is mentioned in common speech, it would seem that the heart plays a comparable part in our conscious experience of human life. But although this organ is said to swell, leap, sink, ache and break, we are aware of its existence only because we have learnt to use such phrases. Admittedly, there are many other feelings that can be shown to arise from the heart – flutters in the chest, pounding in the neck, rhythmic roaring in the ears, throbs, syncopations and unsettling percussions. But these bodily sensations that accompany dread, rage or passion do not convey the impression of a single, localised organ whose disordered action is responsible for such feelings. Once we have been entrusted with this knowledge, however, we tend to report such sensations as if their origin were self-evident.

It is almost impossible to think back to the time when the heart had no place in the collective imagination, when clinical wisdom and anatomical knowledge had not yet organised these feelings into a picture. But there must have been a time when the various diseases that were later shown to be associated with the heart were experienced as individual sensations – a flutter here, a throbbing there, breathlessness perhaps. What reason could anyone have for thinking of these orphan twinges as anything other

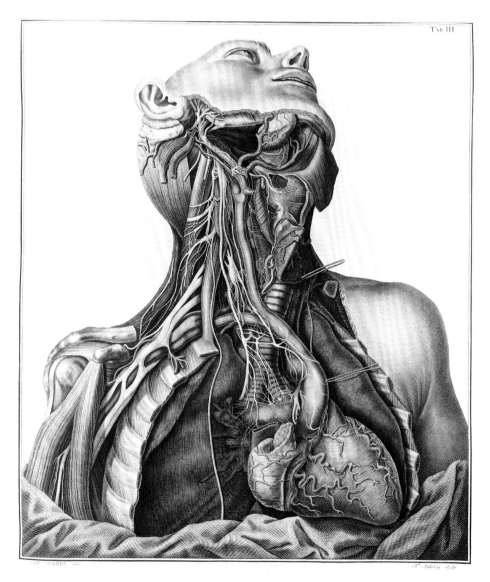

TAB III

1 *Ecorché* drawing of neck and chest, in Antonio Scarpa, *Tabulae nevrologicae*, tab III, published by B Comini, Pavia, 1794, engraving

than what they were? How did the heart make its debut? Presumably when someone first opened the chest and found it there. That was the least that had to happen. But that alone would not have been enough. Taking a look does not automatically reveal what there is to be seen: the innards are not labelled and arranged on shelves, and because we have inherited a clear-cut inventory of the parts we own, it is easy to forget that there was a time when no one knew that there were such parts, and certainly did not know one from another – 'this one' as opposed to 'that one'.

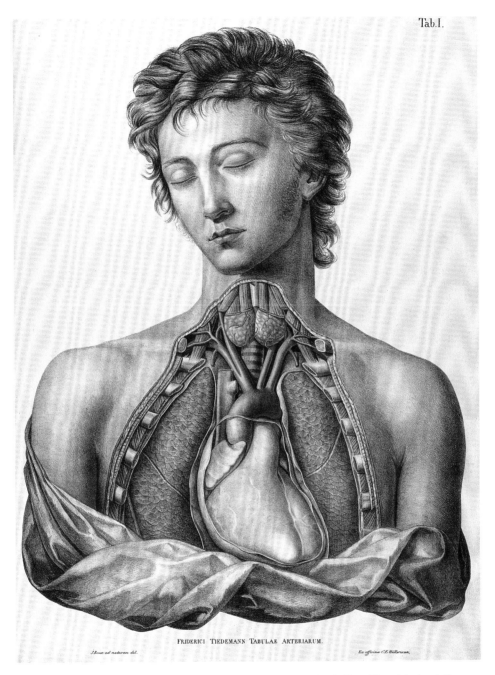

2 Drawing showing partial dissection of the chest of a man, with arteries indicated in red, in Jacob Roux and Friedrich Tiedemann, *Tabulae arteriarum corporis humani*, tab I, published by Officina Christiani Frederici Müller, Karlsruhe, 1822, lithograph with watercolour, 67.5 × 48.8 cm

Nowadays, a demonstrator can open the thorax expecting to find this and that, and, by pointing with his finger, can show what he means. But the visible *thisness* of the heart is not quite so clear-cut as one might imagine. It certainly has distinguishable contours, and abrupt changes of colour – and there would be an extra incentive for regarding that patch of the body as a separate thing if it happened to be independently movable as well. The thing we now call 'the heart' is distinctly beefier than those pink spongey things that seem to fall back from it on either side when the chest is opened. That, at least, is grounds for labelling it 'the heart', and those 'the lungs'. On the other hand, the heart and lungs are tethered to one another by tubes and membranes, and unless you already had a theory that insisted that there was a significant difference between the object you wanted to pluck out of the chest and all those pink tubes that prevented you from doing so, there would be no particular reason for labelling this 'the heart' and those things its 'vessels'.

Anatomical textbooks give the misleading impression that everything in the chest is immediately distinguishable. In the illustrations, the heart is artificially distinguished from its vessels by a bold, graphic outline and sometimes a special colour. The aorta is printed in scarlet; the great veins in sky-blue; the nerves are usually represented in green or yellow. The unsuspecting student plunges into the laboratory carcass expecting to find these neat arrangements repeated in nature, and the blurred confusion that he actually meets often produces a sense of despair. The heart is not nearly so clearly distinguished from its vessels as the textbook implies, and at first sight the vessels are practically indistinguishable from one another. A practised eye can readily recognise the gristly pallor of an artery as opposed to the purple flabbiness of a vein, but what *makes* the eye practised are the theories or presuppositions that direct its gaze – and one of the leading theories of anyone now looking into the chest is the one that says that arteries and veins are different because the blood flows through them in different directions. The colour-codes that decorate students' textbooks are not simply vivid illustrations of what there is to be seen, but graphic conventions that illustrate theories about the function of what there is to be seen. And this, of course, was lacking when men first looked into the human chest.

It is tempting to assume that this visual confusion would have resolved itself as soon as passive inspection gave way to active dissection: that the arrangements would have made themselves clear as soon as anyone began to tease apart and display the matted structures. After all, a skilled dissector can separate and free the various vessels that issue from the base of the heart simply by snipping away the sleeves of connective tissues that bind them all together. But what do we mean by a skilful dissector? If we mean someone who can reveal what there is to be seen, that presupposes that he knows what is there to be revealed: his actions are led by his expectations. What makes him skilful is not the steadiness of his hand, but the knowledge that guides it. Knowing

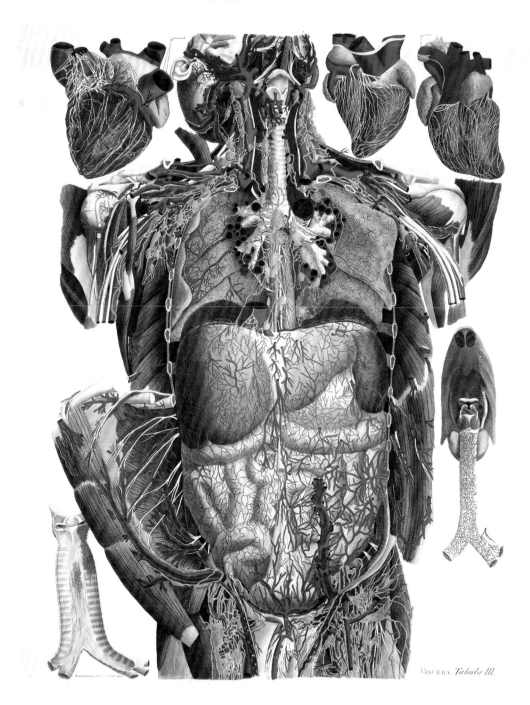

VISCERA *Tabula III.*

3 Anatomical illustration of human viscera in 'exploded' thorax, in Paolo Mascagni, *Anatomie universae*, viscera tab III, published by N Capurro, Pisa, 1823–33, coloured engraving

where to begin implies some previous idea of how one wants to end up. Otherwise, why slice here rather than there?

The difficulties start even before the dissection begins. How would you set about opening the chest to get the best view? What would count as the best view? By now, there is a canonical viewpoint that displays all the known structures to their best advantage and does the least damage on the way in. Any incision is likely to injure some underlying structure, but since the modern investigator knows what is at risk, he can choose the safest incision and, as he already knows which structures he will be sacrificing by going in here rather than there, he can make allowances for what he has destroyed *en route*.

Once inside, the problem becomes even more complicated. Snipping away a sleeve of connective tissue implies that you have already recognised the difference between a trivial covering and the essential structures that lie underneath. Looking at it for the first time, though, you might easily regard a bundle of vessels as a single structure, and the act of unbinding them might seem destructive rather than demonstrative. You cannot know how deeply to cut and at what angle to hold the scissors until you have some knowledge of what you want to show and what there is to damage. The very choice of instruments is an implicit acknowledgement of what you are after and of the risks you are running. When someone favours a blunt probe, it is because he knows the structures that would be endangered by the use of a sharp scalpel. There is no such thing as 'taking care' in the abstract: care and caution are defined by the known risks one is trying to avoid, by knowing what would count as having gone wrong. How could a primitive investigator have known this? Even with textbooks to guide you, it is not always easy to recognise when you have made a mistake. An accidentally cut vessel may gape for a moment, giving some indication that one has made a blunder, but if it is sliced clean across, the two ends may retreat into the ooze and mess of the surrounding tissues, and, unless there were good reasons for pursuing them, these elastic fugitives might get left out of the final count.

Considering all these difficulties, it is amazing that any progress was made, although, given time, persistence and conscientious curiosity, the regularity of the layout would have eventually made itself clear. But that begs another question. Since the first appearances are so discouragingly confusing, why would anyone want to persist? It is only to us that the answer seems obvious: to find out how it all works. But the very idea that 'it' – all that stuff inside – was something that worked or performed is a theoretical assumption that would not necessarily have occurred to the naive observer. It is easy to understand a vague curiosity about one's own interior, a wish to take stock of all those red slippery things that are sometimes revealed as a result of accidents and wounds. But why should anyone think of them as 'the works' as opposed to 'the contents' – as a system rather than an inventory? Why should the notion of 'the works' apply to man at all? Our actions, our perceptions and our liveliness seem self-explanatory. We can do most of the things we want to do, and we can feel, see, hear,

4 J Lamsvelt (Netherlands, b 1674), *Statue of Priapus*, early eighteenth century, engraving

smell and touch all that we need to. We are in business. We are in charge. Events are in hand. Things get done. For the normal unselfconscious person, for the person who is satisfied to get on with it, the idea that there might be something that mediated or mechanised his getting on with it would seem bizarre.

The moment when man first suspected that what he did was the result of hidden things getting done must have changed his whole view of the sort of thing he was. The suspicion that his effectiveness or agency was caused by something other than his conscious urge to be effective, and that he himself had no real control over this, constitutes an almost inconceivable leap of the imagination, and one can only conclude that it was largely the result of drawing an analogy between himself and his own technological artefacts. In primitive societies, where technical images are few and far between and very simple at that, most explanatory metaphors are drawn from nature. In the effort to understand his own make-up, primitive man inevitably resorts to images of wind and water, breezes and tides, floods, fruits and harvests. But the development of technology created a new stock of metaphors – not simply extra metaphors, but ones altogether different in their logical character. Once man succeeded in making equipment that performed – looms, furnaces, forges, kilns, bellows, whistles and irrigation ditches – he was confronted by mechanisms whose success or failure depended on the efficiency of their working parts: things that could block or break, silt up or go out, mechanisms that were intelligibly systematic and systematically intelligible. By mechanising his practical world, man inadvertently paved the way to the mechanisation of his theoretical world.

The success of modern biology is not altogether due to the technology with which we pursue it; the number of technical images we now have for thinking about it play an almost equally important part. An American scientist once said that the steam engine had given more to science than science had to the steam engine, and the same applies to telephone exchanges, automatic gun-turrets, ballistic missiles and computers. Whatever these devices were designed to do, they have incidentally provided conjectural models for explaining the functions of the human body. And the effectiveness of this process grows by geometrical progression. After all butchers, priests and augurers have been disembowelling animals since the dawn of time, and the battlefields of antiquity would have given ample opportunity for looking inside the human body. One of the reasons why the anatomy and physiology of the heart took so long to develop was the lack of satisfactory metaphors for thinking about what was seen. Of course, the written evidence of early Greek science is so fragmentary that it gives an unreliable picture of what was known or believed. Modern scholars continue to disagree about the authorship of these tantalising fragments, and since much of it takes the form of poetry rather than discursive prose it is almost impossible to get a coherent picture of what the sixth-century BC Greeks knew. The organs contained in the chest had already been identified, and the idea that blood ebbed and flowed in the various vessels was also appreciated, but there was no consistent doctrine and no argued conviction about its mechanism. Until the end of the fourth century BC, no one distinguished between arteries and veins, and even then, the distinction didn't carry the systematic significance that it does now.

It would be a mistake to assume that early Greek physiology was as incoherent as the ruined evidence might lead one to believe. Galen inherited a vast treasury of texts and, although most of these are lost to us, the way in which he repeatedly acknowledges the work of his ancient predecessors implies that he was not the first scientist to visualise the blood vessels as part of an intelligible working system. Nevertheless, the weight and cogency of what he had to say is so immeasurably greater than what came before that one is forced to conclude that Galen had some peculiar advantage over his Greek predecessors. And although the unprecedented naturalism of the Roman art of his time indicates that an intense interest in the appearance of the physical world had an important part to play, it is not unreasonable to deduce that this advantage was connected with the technological ingenuity and richness of Roman civilisation.

The exact date of Galen's birth is unknown – perhaps 129 AD – in the Greek city of Pergamum. He received his education in philosophy and medicine first in Smyrna, then in Corinth and finally in Alexandria, where anatomy and physiology had flourished 300 years before. He returned to Pergamum, and became a surgeon to the gladiators – which would have given him a good opportunity to see gaping wounds and flowing blood. He came to Rome in 169 AD and was appointed physician to the pagan emperor Marcus Aurelius. His literary output was enormous: although many of his manuscripts were destroyed by fire in the year 192, his surviving texts amount to more than 9,000 pages devoted to philosophy, medicine and physiology. Much of this consists of careful commentaries on the work of his great predecessors. He believed that all progress took its origin from ancient wisdom, and although he was committed to personal observation and experiment – he insisted on skinning his own animals rather than leaving this menial task to a slave – he always conducted his enquiries in the context of the great tradition that had been established by Hippocrates, Plato and

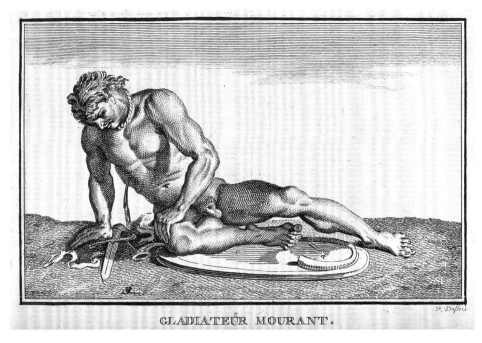

GLADIATEUR MOURANT.

5 Philothée-François Duflos (France, c 1710–1746), *Wounded Gladiator*, eighteenth century, engraving

Aristotle, and by the Alexandrian physicians of the fourth century BC. He inherited and consolidated the traditional notion that the universe was composed of four basic elements, and that these represented the four irreducible qualities of wetness, dryness, warmth and cold. He agreed with the group of writers who are now included under the name 'Hippocrates' that the four universal elements were reproduced in the human body by a quartet of physiological 'humours' (see Chapter 5), and that these took their origin from the elements to be found in food and drink.

For Galen, in fact, physiology started with nutrition. Nutrition was an act of creative transformation that systematically converted the inanimate elements into the active ingredients of the living body. In the first stage, food was absorbed from the gut and passed via the portal vein into the liver, where it was brewed or concocted into blood and at the same time imbued with a weak form of vital principle that Galen called 'Natural Spirits'. On leaving the liver, this raw material ran in an indolent current through the great veins, which distributed it to the hungry tissues. Depleted by the tissues of its Natural Spirits, the blood sluggishly regurgitated back along the same channel, where it was replenished with fresh nourishment by the liver.

Galen recognised that although this theory explained how the body maintained its bulk, some extra principle was required to account for its warmth, verve and spontaneous reactions. The Natural Spirits concocted in the liver might be enough to explain the vegetative functions of the body, but Galen insisted that the blood that went to the brain had to be supercharged with extra force or energy. This, he suggested, was obtained from the air for, like his predecessors, he regarded the invisible element that filled the atmosphere as a vital force.

And this is where the heart enters the system: as the creative mediator between blood and breath. According to Galen, some of the blood leaving the liver overflowed or was diverted into the right side of the heart, where it came into immediate contact, or so he supposed, with air, which had been drawn in through the lungs and along the pulmonary veins. The encounter produced a mysterious incandescence, a biological flame that heated the blood and gave it the fertile warmth that is so characteristic of the living body.

Galen likened the heart to a lamp, fed by the oily fuel supplied from the liver. The smoky fumes exited through the flues of the pulmonary veins and the windpipe – anyone could see that breath steamed or smoked in the cold air. But it was more than a heating system. The blood was not only consumed by the cardiac fire, it was transformed and refined: converted from the thick, purple ooze supplied by the liver into the swift, scarlet stream that issued from the arteries. In this respect, the heart resembled a smelter's furnace, burning off the combustible impurities that were present in the food, leaving a purified residue that was enriched with the peculiar pneumatic principle said to be present in air. The leaping red fountain that shot from the left side of the heart differed from the turbid material entering it from the right, not simply in lacking combustible impurities, but in possessing a weightless substance that Galen called 'Vital Spirits'. This superlative substance was distributed through the arterial system, some of it going to the general tissues, where it presumably reinforced the effect of the Natural Spirits, and some of it going to the brain, where it underwent further refinement and became Animal Spirits – the material responsible for converting thought into action.

This elaborate theory brought together and reconciled the scattered theories of antiquity, organising them into an intelligible system; an industrial plant half-way between a brewery and a blast furnace. It seemed to make sense not only of the various pipes and tubes that were known to exist, but of the various material intakes and outputs of the living body. It explained the source, function and distribution of nourishment, the purpose of breathing, and the spontaneity of the living body. Its dependence on technological metaphors is self-evident: the most notable feature of the system is the emphasis on manufacture and transformation, cooking, brewing and smelting – processes that convert, purify and refine tangible substances. The heart, like the liver, is simply another part of the factory.

The recognition of the propulsive role of the heart was delayed for nearly 1,500 years, although the necessary evidence was just as available to Galen as it was to William Harvey. The difference between the two men is not one of ingenuity and skill – in fact, if these were the sufficient conditions of scientific progress, Galen rather than Harvey might have been the discoverer of the circulation of the blood. But seeing is not all that there is to believing; belief determines the significance of what is seen. The difference between Harvey and Galen was one of metaphorical equipment. When Galen tried to systematise the relationship between the blood and the breath, the co-operation between the liver, the lungs and the heart, the processes on which he modelled his theory were the most conspicuous features of the world in which he lived. There were no better analogies than those of the lamp or the smelter's furnace: they were the most intelligible images of transformation and change. One can only assume that Galen's inability to see the heart as a pump was due to the fact that such machines did not become a significant part of the cultural scene until long after his death. The heart could be seen as a pump only when such engines began to be widely exploited in sixteenth-century mining, fire-fighting and civil engineering.

In the absence of a more plausible metaphor, Galen's industrial model inevitably monopolised the imagination of late antiquity, blinding men to inconsistencies that later became self-evident. For instance, the furnace model required the blood to move directly across the heart from right to left, in spite of the fact that its passage is quite obviously blocked by the thick muscular wall that separates the two ventricles. We now know that the only way for the blood to get from one ventricle to the next is by going the long way round through the lungs, and that the transformation from purple to scarlet takes place there rather than in the heart. Galen's system gave little or no emphasis to the pulmonary circulation, dismissing it as a lubricant trickle that nourished the bellows of the cardiac furnace. Since his theory depended on a direct transit across the heart, he insisted that the septum that divided the two ventricles was perforated by channels. He admitted that these channels were too small to allow the passage of thick, unrefined blood, but since his theory also insisted that the blood was refined by its encounter with air, this did not pose a problem.

This casuistry shows how persuasive a metaphor can be, and how, once an idea lodges in the imagination, it can successfully eliminate or discredit any evidence that might be regarded as contradictory. No special technique was needed to show that the septum is impermeable: although its surface has an irregular, corrugated appearance that might give the impression that it is pitted with the mouths of small vessels, a quick poke with a bristle would have shown that these pits lead nowhere – certainly not from one side to the other. Galen could have performed this experiment just as easily as his successors did. The great sixteenth-century anatomist Vesalius did perform it, but since he, too, was under the spell of the traditional theory, he insisted that the blood 'sweated' across the muscular wall through pores that were admittedly too small to allow the passage of a bristle.

Philosophers of science sometimes imply that scientific thought is a simple alternation between conjecture and refutation, and that contradictory evidence automatically discredits an otherwise plausible hypothesis. The history of cardiac physiology shows that this is an over-simplification, because it overlooks the criteria that are used to decide what will count as a contradictory finding: if a theory has found favour with a scientific community, it is the anomalous finding rather than the theory itself that is discredited. This is exactly what happened in the experiment with the bristle. It was only when the pulmonary circulation was accepted as an established fact that the results of this experiment could be recognised as significant. Meanwhile, as is so often the case, it was more convenient to make an *ad hoc* modification of the existing theory. The history of science is full of such examples. If a theory is persuasive enough – and the reasons for any given theory being so persuasive are often hard to list – scientists will accommodate inconsistent or anomalous findings by decorating the accepted theory with hastily improvised modifications. Even when the theory has become an intellectual slum, perilously propped and patched, the community will not abandon the condemned premises until alternative accommodation has been developed.

The same principle applies to the problem of blood flow in the veins, especially to the valves that are to be found in these vessels. Galen could have seen these structures just as easily as his successors did. You have only to open the veins to see that there are little flaps on the inside wall and, from the way these flaps are facing, their valvular function now seems obvious. To us, as to Harvey, they are the most significant contradiction to Galen's theory that the blood could flow hither and thither in the veins: as Harvey saw, the flaps are arranged so that the blood can flow only one way through them. He proved that the flow was one-way by a dazzlingly simple experiment. He tied a tourniquet round the upper part of his arm, just tight enough to prevent the blood flowing back to the heart through the veins, but not tight enough to prevent blood entering the arm through the arteries. The veins swelled up below the tourniquet and remained empty above it, which implied that the blood could be entering them only through the arteries. By carefully stroking the blood out of a short length of vein, he saw that the vessel filled up only when the blood was allowed to enter it at the end that was furthest away from the heart.

These experiments are so simple that it seems surprising that they weren't performed before. But it would not have occurred to anyone to carry out an experiment like this unless he suspected that there was something wrong with the traditional theory, or that there was an alternative theory whose implications required such an experiment as confirmation. But alternative theories can be conceived only against a background of well-established tradition. One of the reasons why it took so long to overthrow the Galenic theory was not that men were overawed or enslaved by it, but that they were not fully acquainted with it. For more than 1,000 years after Galen, the bulk of Greek scientific literature was either lost, forgotten or neglected, and most of Galen's work remained untranslated until the revival of ancient learning marked the beginning of the Renaissance. A theory that survives uninterruptedly has quite different consequences from one that is revived after a long period of neglect (this also applies to art or literature – imagine how different our attitude to Shakespeare would

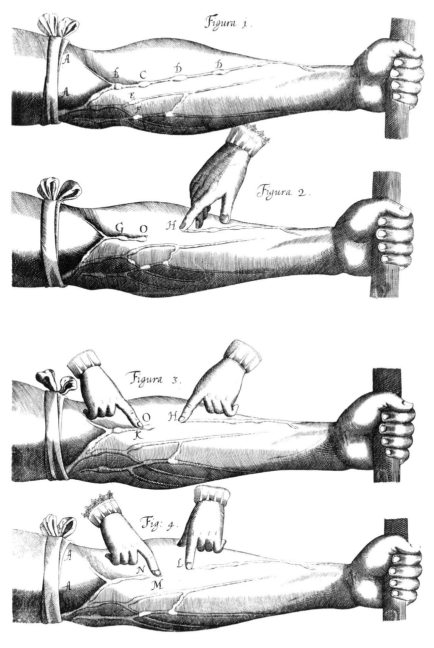

6 Experiments to demonstrate the function of the valves in the veins, in William Harvey, *Exercitatio anatomica de motu cordis*, figs 1–4, published by G Fitzer, Frankfurt, 1628, engraving

be if his plays had been taken out of commission between 1700 and 1900). When theories are vigorously discussed, examined and contradicted for any length of time, they tend to evolve of their own accord. In fact, as long as a theory continues to occupy the public domain, it automatically sets up reactions that lead to its own development and eventual downfall: its uninterrupted existence, its acknowledged claim to be a subject of controversy, guarantees not so much its survival, as its inevitable replacement by more convincing alternatives.

Galenic physiology was never given the opportunity to undergo the ordeal of unremitting criticism, so that when it was re-examined in the sixteenth century it had not yet spent its powers of persuasion. Because of the ascendancy of Christianity, for more than 1,000 years, interest in the nature of the physical world was supplanted by a passionate concern with the metaphysical fate of mankind. The physical order of nature seemed trivial by comparison with what its members had been offered by the redemptive sacrifice of Jesus Christ. The biological origins of the human individual paled into insignificance in the light of what was now known about his theological destiny.

It is one of the paradoxes of scientific history that the great leap forwards took place only when scholars went back to the scientific texts of antiquity. A tension between tradition and fresh observation marks the work of the Belgian anatomist Andreas Vesalius. In 1543, the year that Copernicus issued his work on the revolution of the heavenly bodies, Vesalius published the first modern textbook of anatomy. The illustrations are so beautiful – they were probably done by a student of Titian – that it is easy to overlook the text, and yet it is here that the connection with Galen is recognisable. Medieval textbooks of anatomy have no systematic scheme, whereas Vesalius closely followed Galen's method and reconstructed the human physique from the foundations of its skeleton: 'As poles are to tents and walls are to houses so are bones to living creatures.'

But Vesalius also sets himself the task of surpassing the work of his great predecessor and he undertook to perfect his means of expression by an unflinching study of nature. But although he had the advantage of working with human corpses, where Galen had to make do with barbary apes, in his interpretation of the function of the heart he reproduced all the errors of the ancient wisdom.

In the first edition of his book, he repeated Galen's claim that air entered the left ventricle of the heart, where it served the double purpose of cooling the innate heat and preparing vital spirits. Like Galen, he insisted that the blood soaked plentifully through the inter-ventricular septum, and that, although this wall was one of the thickest parts of the heart, it was perforated throughout by little channels. By the time Vesalius was ready to publish a second edition, he had begun to entertain serious doubts about the existence of such perforations. These doubts may have been reinforced by the emergence of a theory that made such a direct transit unnecessary. Shortly before Vesalius

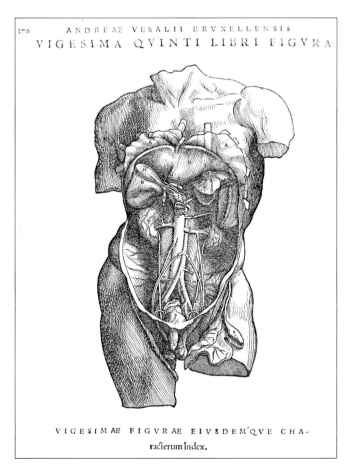

7 Male figure showing viscera, in Andreas Vesalius, *De humani corporis fabrica*, folio 370, fig 20, published by J Oporinus, Basel, 1543, woodcut

published his second edition, the Spanish philosopher-theologian Michael Servetus had suggested that the blood made its way from right to left through a north-west passage in the lungs. This physiological theory appears in a Unitarian treatise for which Servetus was burnt at the stake in Geneva by John Calvin in 1556.

Given the traditional relationship between air and the soul, it is not altogether surprising that a theologian should have concerned himself with the physiology of breathing. Like his predecessors, Servetus recognised that inanimate matter had to be galvanised by some special principle, but he believed that the fertilising encounter between air and blood took place in the lungs, not in the heart. Vesalius probably read the heretical treatise, and that may have been what prompted him to perform his experiment with the bristle. It was left to one of his students to make the conclusive break with the traditional theory.

In 1559, Realdo Colombo published a book entitled *De re anatomica.* In terms of pure anatomy, he had very little to add to the work of his great teacher, but, when it came to understanding the function of the lungs, Colombo was so imaginative that Harvey later copied out the following passage from his book:

> Everyone thinks that there is a way open for the blood to pass from the right ventricle to the left, and that this may be more easily accomplished, they think that it is refined in the transit . . . But they err by a long way, for the blood is carried to the lung through the pulmonary artery and in the lung it is refined, and then together with the air it is brought through the pulmonary vein to the left ventricle of the heart.

Colombo pointed out that Galen's theory not only required blood to pass across an impermeable septum, but also implied the presence of air and smoky fumes in the pulmonary blood vessels. He showed that the pulmonary vessels were full of blood and there was no way for air to pass from the lungs into the heart: the only way was for the blood to visit the air in the lungs.

The publication of these ideas did not automatically lead to the downfall of Galen's theory. There are certain moments in the history of science when the introduction of a valuable truth merely adds to the confusion. At the end of the sixteenth century, physiology was in the same condition in which Copernicus had found astronomy nearly sixty years before: 'It is as if an artist were to gather together the hands, feet, head and other members from many different models, each part excellently drawn, but not related to a single body. And since they in no way match one another, the result is more like a monster than a man.'

Although Harvey was aware of Colombo's dissenting opinion, he was unable to take advantage of it until he had reconsidered some of the other traditional assumptions about the actions of the heart. For example, the official theory claimed that the heart filled itself like a syringe, by actively expanding and drawing the blood up from its natural source in the liver to a level where it spontaneously overflowed into the veins. As the overflow escaped, the heart shrank until it was ready to re-expand, creating a vacuum that would suck in a fresh supply of blood.

This theory takes little or no account of the heart's muscular contraction, and although it was generally accepted that the heartbeat has two phases – swelling followed by shrinking – the traditionalists thought that the heart was only truly active at the moment when it was expanding, and that the contraction was simply a question of elastic recoil. To some extend, this misunderstanding came from the fact that Classical physiologists knew next to nothing about the mechanics of muscular contraction, or even that there was such a thing. If you watch your own biceps contracting, it is easy to get the impression that it is increasing in volume. The fact that

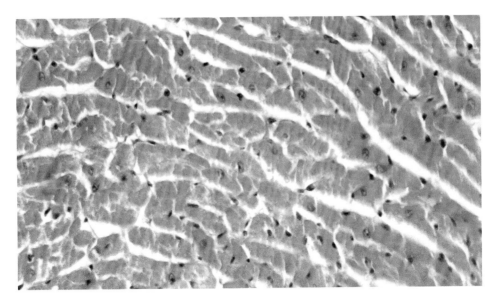

8 Normal cardiac muscle, photomicrograph

such a mistake can be made about a muscle that can be moved slowly at will explains why it was so difficult to identify what was happening in an organ that was beating seventy times a minute. When Harvey first opened the chest of a living mammal, the heartbeat was so rapid that he was unable to tell the difference between expansion and contraction, diastole and systole, let alone identify the active phase:

> I kept finding the matter so truly hard, so beset with difficulties that I all but thought . . . that the heart's movement had been understood by God alone. For I could not rightly distinguish . . . [when] or where constriction and dilation occurred. This was because of the rapidity of the movement, which in many animals remained visible for but the wink of an eye or the length of a lightning flash, so that I thought I was seeing now systole from this side and diastole from that side; now the opposite; the movements now diverse, and now inextricably mixed. Hence my mind was all at sea and I could neither come to a decision myself nor assign definite credit to others.

Recognising that the problem was one of speed, Harvey imaginatively switched his attention to animals whose heartbeat was slow enough to let him see the two phases one after the other: cold-blooded creatures such as toads, serpents, frogs, snails, lobsters and prawns. He also studied the dying hearts of warmer animals, finding that as they began to flag and move more languidly he could inspect and determine the nature of the movement.

From these observations, he concluded that the heart was active not when it was expanding, but when it was most vigorously shrinking, when its walls were thickening. This, he decided, was the moment when the heart-beat made itself visible on the outer wall of the chest. By cutting the arteries leading out of the heart, he discovered that the blood was ejected into them at the moment when the walls thickened and bunched up, and that the blood re-entered through the veins as this active contraction relaxed. He pointed out that the active phase was one of contractile movement and that, contrary to traditional belief, the heart filled itself following the moment of its passive relaxation or collapse. These observations immediately made sense of the arterial pulse, for Harvey recognised that the percussion that could be felt at the wrist coincided with the propulsive thrust that could be detected by feeling the outer wall of the chest. Harvey compared the arterial pulse to the effects of 'blowing into a glove and producing simultaneous increase in volume of all its fingers'.

By working with slow hearts, Harvey was also able to see that not all the parts of the heart contracted simultaneously in a synchronised burst, but that there was an orderly sequence of contractile events, which started in the receptive chambers of the auricles and proceeded towards the propulsive chambers of the ventricles:

Those two movements, one of the auricles and the other of the ventricles, occur successively but so harmoniously and rhythmically that both (appear to) happen together and only one movement can be seen, especially in warmer animals in rapid movement. This is comparable with what happens in machines in which, with one wheel moving another, all seem to be moving at once. It also recalls that mechanical device fitted to firearms in which, on pressure to a trigger, a flint falls and strikes and advances the steel, a spark is evoked and falls upon the powder, the powder is fired and the flame leaps inside and spreads, and the ball flies out and enters the target; all these movements, because of their rapidity, seeming to happen at once as in the wink of an eye. In swallowing too it is similar. The root of the tongue is raised and the mouth compressed and the food or drink is driven into the fauces, the larynx is closed by its muscles and by the epiglottis, the top of the gullet is raised and opened by its muscles just as a sack is raised for filling and opened out for receiving, and the food or drink taken in is pressed down by the transverse muscles and pulled down by the longer ones. Nevertheless, all those movements, made by diverse and opposite organs in harmonious and orderly fashion, appear, while they are occurring, to effect one movement and to play one role which we style 'swallowing'.

By now, Harvey was convinced that the heartbeat was an act of vigorous propulsion, and the implications of this conclusion led to the complete overthrow of the traditional theory. He recognised, for instance, that the amount of blood propelled into the arteries at each beat was such that if it were multiplied by the number of beats in an hour the total volume would be more blood than the body contains. His measurements of the output were relatively inaccurate, but the fact that he was prepared to see the problem in quantitative terms at all marks him as the first modern biologist. A more pedantic man might have got bogged down in finicky measurements but, like all great scientific geniuses, Harvey saw the implications of even a rough approximation. He recognised that the hourly output of the heart was so large that the only way it could replenish itself was by taking in through its back door all that it had thrown out through the front: 'In consequence, I began privately to consider if it had a movement, as it were, in a circle.' It is hard to say when or how Harvey first had this insight. He had been investigating the subject for at least ten years before he published his great book in 1628. Towards the end of his life, he told Robert Boyle that the idea of circulation first occurred to him as a result of seeing valves in the veins:

> In the only Discourse I had with him . . . [he said that] when he took notice that the Valves in the Veins of so many several parts of the Body, were so Placed that they gave free passage to the Blood Towards the Heart, but oppos'd the passage of the Venal Blood the contrary way: he was invited to imagine that so Provident a Cause as Nature had not so Plac'd so many valves without Design: and no Design seem'd more probable, than That, since the Blood could not well, because of the interposing Valves, be Sent by the Veins to the Limbs, it should be sent through the Arteries, and return through the veins, whose valves did not oppose its course that way.

But the way in which a scientist remembers and publishes his arguments is not necessarily the order in which the idea originally occurred to him, and since Harvey did not log his daily thoughts, we shall never know for certain how the notion of a continuous circulation first occurred to him. Scientists are notoriously forgetful about the origin of their most interesting conjectures, and although the existence of valves confirmed Harvey's hunch that the blood flowed one way through the veins, it seems likely that he had already recognised the existence of a one-way circulation, and that this made him realise that the flaps on the inside of the veins *had* to be valves.

As far as one can tell, his most fruitful insight was his recognition of the propulsive power of the heart, coupled with the experiments that confirmed it. By cutting arteries, Harvey showed that the rhythmic spout of blood invariably issued from the end nearest to the heart and that it coincided with the moment when the heart contracted and whitened. By tying ligatures at strategic points throughout the circulatory system, he showed that the vessels became empty and pulseless beyond the block, swollen and engorged with blood behind it.

As already pointed out, Harvey would not have been prompted to perform such experiments unless he had already had a hypothesis that forecast their outcome. How did he first entertain an idea that so systematically contradicted all that had been suggested previously? His personal observations of the heart's actions played an indispensable role, and it was certainly a stroke of genius to think of examining slow hearts. But once again the influence of technological metaphor must not be overlooked. By the end of the sixteenth century, mechanical pumps were a significant part of the developing technology of Western Europe. Coal and metal mines were being deepened to supply the needs of growing cities. Engineers were bedevilled by the problems of seepage, and forceful pumps were the only way of keeping the shafts empty. Contemporary handbooks of metallurgy included pages of pumping mechanisms. The hydrostatic principles were also applied to the design of ornamental fountains, and in 1615 Salomon de Caus published *Les Raisons de forces mouvantes*, describing a machine for putting out fires:

> The said pump is easily understood: there are two valves within it, one below to open when the handle is lifted up and to shut when it is down, and another to open to let out the water; and at the end of the said machine there is a man who holds the copper pipe, turning it from side to side to the place where the fire shall be.

Historians still disagree about the influence of the fire-pump, but it seems unlikely that Harvey would have departed so radically from the traditional theory if the technological images of propulsion had not encouraged him to think along such lines.

Harvey marshalled all his arguments into a single book of 100 pages. It was originally published in Latin, but even in English translation it has the force of great literature. What strikes the modern reader is the inexorable march of its reasoning, the simplicity of its conjectures and its unflinching determination to develop all its implications in an experimentally checkable form. Naturally, the work is incomplete. It was left to Harvey's immediate successors to find out why the blood circulated through the lungs, but by proving so conclusively that it did he created a soluble problem – which is the least that any scientist can demand of one of his colleagues.

He failed to find the tiny vessels that linked the arteries to the veins, and it was only when the Italian microscopist Marcello Malpighi (1628–94) turned his lens on to the lungs forty years later that the existence of the capillaries was even suspected. If anyone before Harvey had discovered these vessels they would have found it impossible to explain their function, and when the function of something is not recognised, its visible appearance is often misrepresented as well. Things tend to look like what we know they are for – and if we don't know what they do, we often find it hard to say how they look. Harvey might have discovered these vessels for himself if he had possessed a microscope: his theory demanded their presence, and it is a characteristic feature of fruitful scientific theories that they suggest the existence of objects or processes that may not be discovered until the appropriate instruments reveal their presence.

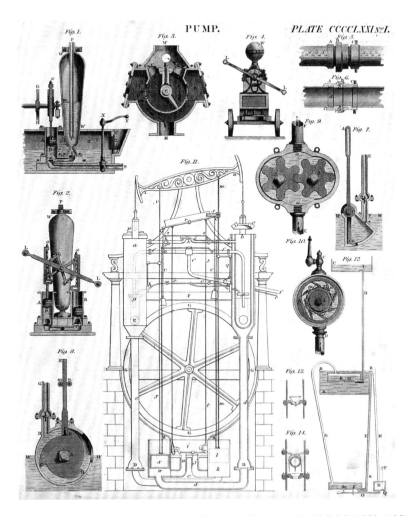

9 Various pumps and steam engine, in John Farey, *Edinburgh Encyclopedia*, Vol II, folio 1332, published by William Blackwood, Edinburgh, c 1813, engraving by James Moffat

Like all good theories, Harvey's bristled with unfinished business, and the fact that he was unable to finish the business himself does not diminish his greatness. All subsequent investigations have been framed by his basic assumption, and, although Harvey would have been puzzled by the sophisticated electronic and biochemical methods now used to pursue these investigations, he would certainly have understood their significance. He might have been mystified by the anaesthetic techniques that enable a modern surgeon to enter the chest and open the heart, but his own theory dictates the replacement of a damaged valve by an artificial substitute: since he explained the function of the walls that separate the various chambers of the heart, he would have understood and applauded the repair of congenital holes.

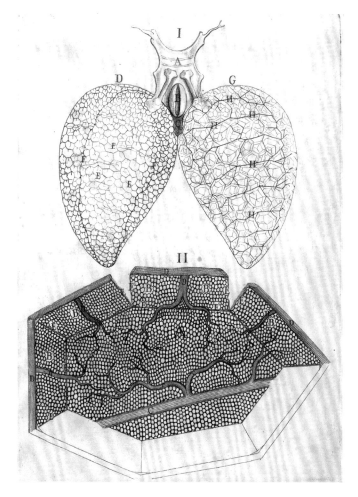

10 Lungs and capillaries, in Marcello Malpighi, *De pulmonibus observationes anatomicae*, published by GB Ferroni, Bologna, 1661, engraving

It would nevertheless be an exaggeration to claim that Harvey's theory foresaw or even implied *all* that we now know about the heart and circulation. It would be too much to expect any theory to do that. Harvey's hypothesis accommodates but does not actually predict what we now know about the circulation. Harvey recognised that the heart could alter its performance, that it could speed up with exertion and change its rate under the influence of strong emotions, but he did not suspect that such alterations were part of an elaborate homeostatic repertoire, or that the heart and blood vessels, like the lungs, kidneys and every other system in the body, are constantly adjusting their behaviour in order to meet the varying stresses of life. He did not know that the arteries are sleeved with muscular tissue and can, therefore, change their calibre and adjust the volume of the circulation to compensate for blood loss, or that the flow can be redistributed both among and within the various organs of the body. It wasn't until the end of the nineteenth century that physiologists discovered the existence of a so-called vaso-motor system, an extensive network of automatic nerves co-ordinated within the brain that regulates the behaviour of the heart and blood vessels.

Harvey's failure to recognise the existence of such mechanisms, and the failure of his theory to predict their existence, are forgivable errors of omission. The function of the voluntary nervous system was not yet understood, so how could he possibly have suspected or even described the functions of involuntary nerves? And since the idea of self-regulation or homeostasis was not to be born for more than 200 years, his failure to appreciate the versatility of the circulation is quite understandable.

11 Chicken and egg, in Fabricius ab Acquapendente, *De formatione ovi et pulli*, facing folio 62, plate 1, published by A Benci, Padua, 1621, engraving

The only point where it could be said that Harvey made a positive mistake was in explaining the physiological origin of the heartbeat itself. He claimed that the action of the heart was not spontaneous, but rather arose from the origin of all spontaneity – the living blood. In his investigations of a developing chick, he detected a pulsing spot of blood whose rhythmic flashes preceded the recognisable appearance of the heart. If his lenses had been better, he would have seen that the pulsation was caused not by the blood, but by the transparent heart which enveloped it. His error was not entirely due to faulty observation. Like his colleagues and predecessors, he was susceptible to ancient dogma, and, although he was able to take an altogether mechanical view of the function of the heart, he could not shake off the traditional belief in the existence of vital principles. In the seventeenth century, the spontaneity or 'go' of living things could be explained only by referring back to some spiritual prime mover, and in Harvey's case this energetic agent was not merely to be found in the blood, it was identical with it.

This is an edited version of an essay entitled 'The Pump', first published in *The Body in Question*, 1978.

Peter Allinson

Telecommunications manager, in conversation with Melissa Larner

MELISSA LARNER: When did you have your heart transplant?

PETER ALLINSON: In 2001. I caught a viral infection in 2000, which led to dilated cardiomyopathy, where the heart muscle is damaged and loses the ability to pump blood.

ML: Do you know why that happened?

PA: No. I've never been able to find out why. They think it's either that it was something I'd recently contracted, or that it had been in the blood for some time, and had lain dormant for a while, but had suddenly activated. They don't really know.

ML: And they told you the only way to deal with it would be to do a complete transplant?

PA: At first, the doctors said, 'We'll try you on a course of medication and see what happens.' This was in late October 2000. In January 2001, I went back for reassessment and they said, 'The medication isn't working at all. You've actually gone downhill rather than picked up again.' So the only way forwards for me was a complete transplant.

ML: During this time, were you very poorly? Were you able to function?

PA: I could function to a certain degree. Towards December, I could barely walk up stairs; I'd have to have a rest at the top to catch my breath. My son was nine months old at the time, and I could pick him up and play with him for a minute, two minutes at tops, and then I'd have to sit down for a rest. I could go out for walks, but I could only get down to the end of the road and back before I was completely out of breath.

Ben Edwards, *Machines 1* (closing up the sternum), 1996–9, C-type print, 244 × 122 cm

ML: And you were only thirty-one at the time. Were you an active person before you became ill?

PA: Yes. I was competing in triathlons. That year, I'd competed in about six triathlons, and I was about to compete in the winter half-marathon, which I did every year. I had to miss that one!

ML: Was there a long wait until they found a suitable donor?

PA: No. But my wife and I thought we were ready for it. Funnily enough, we'd thought about transplantation when we'd first gone to the hospital. We'd talked to a couple of trans-plantees, who were in for MOTs and that sort of thing, and we decided that probably would be the best way to go, because I'd been so active before, and I probably wouldn't be happy until I was active again. And as soon as they got the results back in January, they said, 'Right, you're going to have to have a full transplant.' So we thought we were prepared for it. But we weren't expecting the transplant to take place two weeks later!

ML: So you were lucky, a heart just came up?

PA: Yes. One Sunday evening, we got a call from Harefield. I thought it was my mum because she usually rang on a Sunday night at nine o'clock. But it was the hospital co-ordinators, who said, 'We've got a heart suitable for you. Make your way in.'

ML: What were your feelings then?

PA: I was quite looking forward to it. There was a bit of anticipation on my part, because I'd thought about it for a long time, and now suddenly, it was all starting to happen. And you worry about what's going to happen – will you pull through the operation or not? It's one of those things: you don't know whether you're going to pull through. My wife was a bit more nervous than I was. She was running up and down the stairs. We had to get my brother-in-law to come in and look after our young son while we went to the hospital.

ML: It must have been very frightening.

PA: It was a bit surreal, actually. We put the radio on when we got into the car, and it was strange because we were just singing along to every song on the radio while I drove to the hospital. And because it was a cold night in January, the mist was rolling in. It was just strange. I got into the hospital at about half-past nine, and they started prep-ping me at about three o'clock in the morning. Around about four o'clock, I got into the operating theatre. I said to the anaesthetist, 'Before you put us under, what's it like outside now?' We hadn't seen anything since nine o'clock that night. He said, 'It's really, really foggy now.' We were having a little chat, and he was cracking a few jokes and stuff like that. So it was different from what you expect. You expect everyone to have masks on, and to be very quiet and solemn. But we were being jolly and cracking jokes about the weather and talking about where everyone had come in from. It was a bit of an awkward hour in the morning to have to come in to work.

ML: How did you find your relationship with your surgeon?

PA: I never really talked to the surgeon much. You see consultants. The only time you see the surgeon is straight after the transplant, to make sure that everything's going well. But I send him a Christmas card every year to say 'Thank you very much for the help you've given us.' We're grateful to all the hospital staff because they're so good. Especially the intensive-care nurses, because they see the worst of everything, and then as soon as you go out of their ward, they don't ever see you again. So it's quite nice that every now and again, when you go in to talk to the doctors, you can say hello to the nurses – as well as saying thank you.

ML: So you have to go back to the hospital regularly?

PA: I go back once a month for blood checks. I've got a pacemaker as well – that was put in this year – so I have to go back regularly for check-ups for that, usually every three to six months.

ML: Do you know whose heart you got?

PA: Yes, funnily enough, a couple of months after the transplant I was in the transplant clinic waiting to see the doctor. I'd had to go on steroids because I'd had a rejection a couple of weeks after leaving the hospital. And I started chatting away to the man sitting next to me, who was in the same heart-transplant club as me. We were talking about a meeting that was coming up, which we were quite looking forward to. He was saying that he had a nice pool, and I was saying, 'I can't swim yet because it hasn't been a year post-op.' And he said, 'Oh, I haven't been a year, either.' He started saying things like, 'When was your operation?', 'Oh mine was then as well.' 'Who did your operation?' And I realised I'd managed to sit next to my donor. We'd had a domino operation. He got a heart and lungs from someone who'd had a fatal accident and he donated his heart to someone who needed it, and that just happened to be me! It was completely by accident that we were sitting in the waiting room at the same time, waiting for two different doctors. We knew that only two heart operations had been done at the time – I'd had one and he'd had the other – so because we'd had our transplants on the same day, it was more or less cut and dry that he was my donor. We were both quiet for a minute or two, and then it suddenly just sort of dawned on our faces and James said, 'You've got my heart!' That was it in a sentence.

ML: That must have been an extraordinary moment. So James's heart was fine, but he needed a lung transplant?

PA: James had cystic fibrosis, which is a genetic defect. Progressively through the years, people with cystic fibrosis tend to be OK up until round about their teens, early twenties, and then suddenly one infection can just knock them sideways and they'll never recover from it. So to get over that, they now have lung transplants, which they've perfected over the last couple of years, but when James and I were ill, it was deemed better to have a heart and lung transplant rather than just lungs, because the chances of rejection and possible complications were lessened.

ML: You must have felt an incredible bond with him when you suddenly realised it was his heart that you had.

PA: It was quite a good sensation, actually, because to see someone who's donated his heart to you, up and well and fighting fit and carrying on with his activities . . . It was just really nice to see that he'd pulled through and was still around.

ML: And did you have a sensation that there was part of someone else inside your body or that it had changed your personality in any way?

PA: No. You don't really feel like you've had a heart transplant. Obviously, straight after the transplant, you might, but you've got your sedatives and everything else. And a couple of months down the line, you don't really think about it. Even now, my wife sometimes looks at me strangely and whispers: 'You've had a heart transplant.' Every now and again, you take a step back and remember that you've had a heart transplant. James and I have been asked before about this idea of cellular memory, but we've never had anything like that.

ML: How is your health now?

PA: After the transplant, because I'd spent quite a lot of time in intensive care, I had to get my strength back in my legs and things. But now I'm swimming competitively in a massive Amateur Swimming Association event. I also compete in the British Games and World Games.

ML: That's fantastic. Do you find that it's more exhausting than prior to your illness?

PA: No, I'm really striving to get back to the way I used to be, and at the moment I'm doing quite well. I've put myself down now for the Great North Run next year, so I'm trying to train for that.

ML: How many miles is that?

PA: It's a half-marathon, so it's 13 miles and 150 odd yards, but after 13 miles I don't think you care about the extra yards!

ML: You certainly seem to be taking it all in your stride.

PA: I think from the outset, my wife and I have been like that. We tend to be like that anyway – anything that comes in, we'll take it and do something with it. From being ill to having the transplant was only about four months, so in that time you don't have time to think about things; you just deal with it and get on.

ML: Would you say that the experience has enhanced your life in any way?

PA: I think I appreciate life a lot more now. If opportunities crop up, I grasp them and do something with them. In the past, it would be, 'Oh no, I won't do it this time, I'll do it next year or six months down the line.' I do things spontaneously now. I think about today, rather than what will happen next week or the week after that.

ML: What advice would you give to someone with a similar condition or experience?

PA: Explore all possibilities. If it comes down to having a transplant, then take the transplant. If you can make it better with medication, take the medication. Science is progressing in leaps and bounds now, so within a couple of years they'll be doing things that we haven't even thought of.

The Renaissance Heart
The Drawings of Leonardo da Vinci

Francis Wells

A plaster mould to be blown with thin glass inside and then break it from head to foot . . . But first pour wax into this valve of a Bull's heart so that you may see the true shape of this valve.

Leonardo da Vinci

In this short experimental instruction, Leonardo da Vinci (1452–1519) sets himself apart from all other anatomists of his era and for some time to come. His words reveal a truly enquiring mind that perceives the most important questions to be asked and sets out to find the answers using his unique powers of vision, observations of the natural world, incisive logical deduction and lateral thought allied to embryonic modern scientific methods. His remarkable descriptions of the functional anatomy of the aortic valve have stood the test of time. Other observations and deductions about the structure and function of the heart can still inform current cardiological debate.

Before looking at some of Leonardo's original contributions to the knowledge of the heart's structure and function, it is important to describe the context in which he lived and the state of anatomical knowledge at the time. During the period of history that we have come to call the Renaissance, anatomy stood at the crossroads between science and art. In many ways, it inhabited the middle ground at a time when the distinction between the disciplines that we make today did not exist. Indeed, in that period, the philosophy behind both was a common denominator. The body of man was seen by many as the host of the spiritual soul and hence inviolable. By others, its wonders were seen as part of the macrocosm/microcosm continuum, in which man is a small and integral part of the wider natural universe. This concept, first discussed by Galen, was championed by

1 Leonardo da Vinci (Italy, 1452–1519), sketch of an old man morphing into a tree (detail of page showing geometrical figures), c 1488, pen and ink, 32 × 44.6 cm

Leonardo in his notebooks, and was an expression of the new Humanism of the time, a revival of the style and traditions of the Greco-Roman culture. The separate discipline of scientific method had yet to develop, and philosophy, rhetoric and logic predominated in academic circles. Artists studied anatomy in order to understand the interstices of the human frame so that they could reveal more effectively its power, fragility and reality in artistic form. Physicians relied upon artists to illustrate their manuscripts, such as they were, and both would have met on the premises of merchants who sold materials common to these practices. This common ground was significant.

At that time, the practice of medicine was concerned with remedies for specific complaints, handed down from generation to generation. Surgery was crude, and dealt mainly with the drainage of abscesses as well as domestic and battlefield injuries. Scientific enquiry was not interwoven with medical advances in the way that we expect now, although analysis of the inner workings of the body gathered pace as access to dissection improved (for reasons that will be discussed later). Hypothesis, experimental design and proof were not the order of the day. In the tradition of the Ancient Greek physician Hippocrates, observation of the effects of an intervention did influence decisions with regard to further treatment, but did not extend to therapeutic evolution based upon the sound physiological understanding of organ and system function. The inseparable conjunction of bodily structure and function so characteristic of the working methods of Leonardo was what was to set him apart from all others at that time, even beyond the great Flemish anatomist Vesalius, whose *De humanis corporis fabrica*, published twenty-four years after the death of Leonardo, focused principally on structure and descriptive anatomy, not on function.

During the early years of the Renaissance, anatomical examination by artists in their workshops and studios was tuned to an understanding of the underlying musculature and bony structures for representative purposes rather than for functional knowledge of the body as a physiological integrated unit. In the first century BC, the Roman engineer and architect Vitruvius had discussed in his ten-volume treatise *De architectura* the proportions of the human form as central to his architecture. Working in this tradition, the polymath Leon Battista Alberti (1404–72) suggested three components of anatomical study for application to painting. The first was to know the arrangement of the bones within the human frame, the second was to understand the distribution and arrangement of the muscles, and the third, how to depict the covering layer of the skin, and the flesh that was to be found underlying it. Many but not all of the major studios would have embarked upon dissection of the musculature, beginning with the technique of flaying to reveal the covered world of the muscles and veins. These studies were often represented in the form of *écorchés*, figures rendered in plaster and cast in bronze. Antonio del Pollaiuolo (1432–98), Leonardo's teacher Andrea del Verrocchio (1435–88), Luca Signorelli (1442–1524), Titian (1477–1576), Raphael (1483–1520) and Michelangelo (1475–1564) were all 'anatomical artists', but none left behind reports of their activities, and the evidence of their level of interest as witnessed in their work reveals only a desire to understand form rather than function. Study of the flayed body is perhaps best revealed in Pollaiuolo's famous *Battle of the Ten Nudes* (c 1470).

Leonardo disparagingly referred to Michelangelo's nudes as 'Nutcracker men', due to their inflated musculature, and pleaded with artists through his writing not to exaggerate the musculature to the exclusion of the portrayal of what we would now call 'body language' to convey emotion. Leonardo's earliest remaining anatomical study is of a male figure with the major blood vessels and main organ systems laid out within the frame of the body so as to reveal their connections and relationships. This sketch, executed in the mid-1480s, is among the first, if not actually *the* first, representation of recognisable 'surface anatomy' as we would call it today. Manuscripts in the Bodleian Library in Oxford show a similar display, but with anatomy that is unrecognisable in comparison to that of Leonardo. In this drawing, Leonardo sets out his stall as a different type of artist/anatomist. He is clearly interested in the complex inner workings

2 Antonio del Pollaiuolo (Italy, 1432–98), *Battle of the Ten Nudes*, c 1470, engraving, 38.4 × 59.1 cm

of the body in the way that physician/anatomists were expected to be. The drawing perhaps summarises his knowledge of the heart and its relationship to the liver and the rest of the circulation, which, as we shall see later, is strongly influenced by Galen.

In the fifteenth century, Galen and the Greek philosopher Aristotle were the giants of the accepted science and art of medicine, mostly revealed through Arabic manuscripts and through the translations and interpretation of the Persian physician and philosoper Ibn Sina, known as Avicenna, and the Italian professor of medicine, Mondino dei Luzzi. All of this was tempered with early forms of Humanism. Other than within the work of Galen, it is difficult to find attempts to describe organ function except in the emotive terms expressed through the accepted Hippocratic ideas of the 'Vital Spirits' (see Chapter 5). Furthermore, the matter of all living creatures was thought to be energised and brought to life by a vital force, the *pneuma* or *anima*, seen as the ingredient that differentiated inanimate clay from a living, breathing being. Some modern authorities have equated this with what we recognise today as the essential element of life: oxygen. Since this element could be nothing more than a concept at that time, its mode of action was to remain obscure until it was discovered in 1772 by the Swedish chemist Carl Scheele (1742–86), and almost simultaneously by Joseph Priestley (1733–1804) in England. In 1775, Antoine Lavoisier (1743–94) in France established oxygen as the true element of life.

Into this setting came Leonardo da Vinci: artist, polymath, musician and gifted communicator. The circumstances of his early life were to have a significant impact on his intellectual and artistic development. He was born in or near Vinci, a small town in the hills north of the winding Arno river, on the way from Florence to Pisa. His father Ser Piero and his mother Caterina were not married. At the time, Ser Piero was in his early twenties, perhaps finishing his training as a notary, and Caterina probably in her late teens. Ser Piero was born of a line of Florentine notaries. Caterina was in all likelihood the daughter of a local farm worker. Her family is thought to have lived in or near Anchiano, a hamlet situated a mile or so up the hill from Vinci. Little else is known of Caterina, other than the fact that sometime later she married a local herdsman known colloquially as 'the Quarreler', or Accattabriga di Piero del Vacca de Vinci. This inauspicious start in life did not completely disadvantage Leonardo. Perhaps cared for by Caterina for a year or two, he entered the house of his grandfather, Antonio da Vinci, who along with his uncle was the principal figure in his life in Vinci. It was Antonio who recorded the birth on 15 April 1452, and the baptism the following day, on the back of an old notebook. Documented there were also the births and baptism of his own four children.

The baptism was witnessed by ten godparents in the church of Santa Croce in Vinci. Hence, despite his illegitimacy, Leonardo was welcomed into the world in fine style, surrounded by robust family support. This is important when putting his early life into context. There is a temptation to think that an illegitimate child would have been neglected, as in other cultures and at other times. However, the evidence suggests that he received at least a good basic education, though lacking in some of the Classical gilding that one would expect of a higher-born individual. Throughout his own written record in later life, he reports efforts to come to terms with Latin, Greek and advanced mathematics, subjects that even with his giant intellect he failed to master and for which he relied upon others when knowledge of them become relevant to his endeavours.

His father's profession was a highly respectable one, the notary being an essential functionary who oiled the wheels between lawyers, accountants and investment brokers. The Guild of Notaries in Florence was among the most esteemed of the seven major guilds. As the Medici empire grew, it is likely that Leonardo's father would have grown in stature too. Ser Piero had left Vinci for Florence by 1446. Through records accredited by him, appearing in Pistoia and Pisa, it is clear that as a young and newly qualified notary he moved easily around the region before settling in Florence, where he built a significant career.

Leonardo's uncle, Francesco, was only fifteen when Leonardo was born, and must have had a major influence on his early life, since he never left Vinci. With Ser Piero in Florence, it is most likely that Francesco became a surrogate father to the boy. This assumption is born out to some extent by the fact that Ser Piero left nothing to Leonardo in his will, while Francesco left him everything. This gave rise to a bitter struggle between Leonardo and his stepbrothers over Francesco's will. Francesco tended the family farmland and vineyards, and may well have been the progenitor of the young Leonardo's interest in nature. Certainly, it is clear from his own record that an insatiable desire to understand the workings of nature was Leonardo's driving force from the outset. It was this enquiring mind that would cause him to begin to ask questions about the human form in a way that had been rare up until that time. A relentless persistence in relating form to function, unfettered by preconceived ideas, was to become his hallmark. Leonardo held the belief that every structure in nature is as it is for a reason, and that this reason could be found by describing the forces acting upon it and the function that the structure had to perform. This teleological thought led him to some of the most original experiments that we would now call 'scientific', in his or any time. Free from established learning excavated from the books of others, he trained himself to observe and record what he saw with his own hand, truthfully and with great beauty. The accuracy of his notes allows an intimate dialogue with his work even now, in the twenty-first century.

Leonardo's endeavours covered the arts (painting, sculpture, music and architecture), military engineering, hydrology, mathematics and early physics – in the form of the laws of motion and flight – town-planning and an early form of public health, as well as events-organisation for the nobility. Among all of this, he found time to carry out in-depth studies of anatomy, both functional and applied. At first, as was the case with his contemporaries, it is likely that his work was an attempt to understand the body in more detail for the purposes of artistic representation. This level of involvement would have been instigated in the workshop of Verrocchio, where Leonardo was apprenticed by his father. At some stage, however, it developed into a profound desire to link structure to function, rapidly straying into the realms of physiological research and morbid anatomy.

The apotheosis of this anatomical work rests in the research that he carried out into the workings of the heart. This magnificent achievement was brought about by the integration of many things. First, his indefatigable desire to know and understand how this 'prime mover' worked. Second, his powers of observation and deductive reasoning. Third, his wide knowledge and experience of engineering, hydrology and physical principles, many of which would not be dissected and defined by others for some centuries to come.

Attempts to bring Leonardo's notebooks into chronological order would suggest that there were perhaps three periods during which he devoted himself to intensive activity in the discipline of anatomy. However, it is clear that the interest rarely if ever left him, and we can surmise that his observations spanned a full fifty or more years. The first phase is thought to be the early 1480s through to the mid-to-late 1490s, with a concentrated period in Milan in the mid-1490s. The second is from approximately 1503 to 1508/9 in both Florence and Milan. The third period centres on his sojourn in the Vatican as the guest of Pope Leo X from 1513.

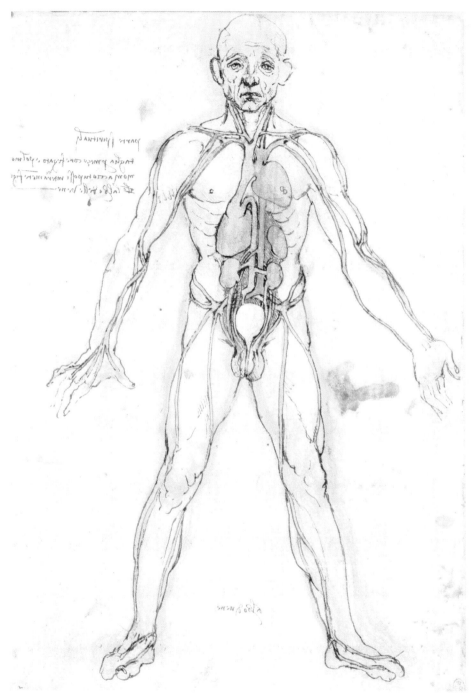

3 Leonardo da Vinci (Italy, 1452–1519), anatomical figure showing heart, liver and main arteries, c 1494/5, pen and ink over black chalk with coloured washes, 28 × 19.8 cm

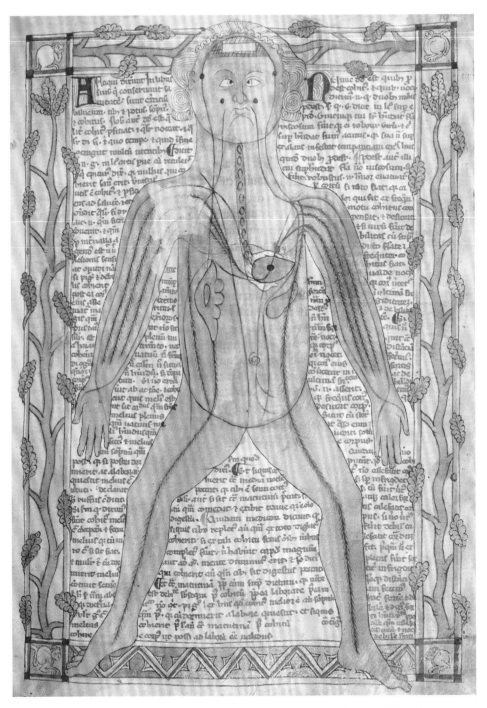

4 Artery Man, in a Latin medical miscellany, MS Ashmole 399, folio 19r, England, late thirteenth century, 26.9 × 19.2 cm

During the first period, Leonardo made some of the most exquisite drawings of the human skull ever executed. These representations of skull dissections were done at around the same time (perhaps as early as 1482) as the drawing of a man revealing the heart, lungs and main arteries referred to earlier. This reveals something of his schizophrenic approach at that early time. Whereas the skull drawings are clearly made from direct observation, the summative drawing is derived from received knowledge rather than his own observation. We know this because the anatomy that is displayed places the liver and not the heart as the source of the vasculature (the arrangement of the blood vessels in the body), a Galenic notion. At that stage in his life, Leonardo was without doubt well aware of, and influenced by, the historical record, and yet when left to his own devices was capable of so much more. His contacts with anatomists and medics through the studio of Verrocchio would have made him aware of contemporary knowledge. He

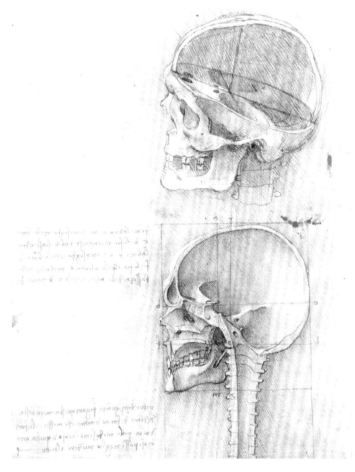

5 Leonardo da Vinci (Italy, 1452–1519), the skull sectioned, c 1506–8, pen and ink, 19 × 13.3 cm

lists in his own library *De medicina* by Aulus Cornelius Celsus (25BC–50AD), as well as works by Aristotle, and quotes from Mondino and Avicenna. However, his attempts to understand vision at that time were revolutionary, and were perhaps the magnet that drew him into the desire to understand and to link structure to function, a hallmark of all of the rest of his work. In the notebook known as *Codex Atlanticus*, the following quotation can be found: 'Since the eye is the window of the soul, the latter is always in fear of being deprived of it.' His studies of vision and his attempts to understand the cerebral integration of the visualised subject are lessons in lateral thinking and the use of parallel science. There is no significant reference to the heart at this time, however.

Dissection in the age of Leonardo was not easily achieved. Although several of the important studios would have access to human remains, the facility with which full dissection could be carried out was variable. Human dissection was not a new phenomenon in the Renaissance: evidence of its practice stretches back to Egyptian times. In his *History of Melancholy*, Democrates reports that the Greek philosopher Hippocrates stated, 'I do anatomise and cut up these poor beasts to see the cause of their distempers, vanities and follies which are the burden of all creatures.' However, as this demonstrates, the dissection process was more in the nature of an autopsy than an attempt to discern structure related to function in the normal setting. Dissection continued in this manner for centuries and, other than in the flaying of bodies by artists, this remained the case until Leonardo began his monumental work. Dissection relied upon the permission of the Church, and was usually carried out in the hospitals associated with the city churches. Famous among them were Santa Maria Nuova in Florence and Santo Spirito in Florence and Rome, the latter within the Vatican walls and used by both Leonardo and Michelangelo. However, the fragility of the relationship between church and practitioner is well demonstrated by the abrupt termination of Leonardo's work in the dissection room in Rome. During his last period of anatomy there, he incurred the wrath of the Medici pope, Giovanni, who became Pope Leo X. A dispute between Leonardo and a German mirror-maker, brought to the Vatican by Pope Leo, resulted in the termination of his free access to dissection.

The Church had other implications for the work of Leonardo, who probably had an uneasy relationship with this institution. To religious leaders of the time, the body was the temporary home of the soul and a vehicle for the Holy Spirit. For Leonardo, however, the body of man was a machine, a vehicle in which to get around and survive. As such, all of its workings needed to be understood. Indeed, in his middle period of anatomy, Leonardo strove to understand the workings of the muscles, bones and ligaments, rather in the way that we might design a robot today. Once he had grasped the relationship of the muscles to the bones, he began to see them in a very modern way: as a system of levers and pulleys, and he drew analogies with his mechanical engineering work, sometimes reducing the muscles to simple cords to understand their actions. To many in the Church, this would have seemed sacrilegious.

The other factor that made dissection difficult, and indeed quite unpalatable for some, would have been the awful sight of human remains pulled apart in the finest detail, along with the stench of decomposition. Leonardo writes graphically about this in his notebooks:

> And if you have love for such things [dissection] you will perhaps be prevented by your stomach, and if this does not prevent you, you may perhaps be prevented by the fear of passing the night-time in company with bodies flayed and fearful to look upon. And if this does not prevent you, perhaps you will lack good draughtsmanship that should belong to such drawing, and if you have the draughtsmanship it may not be accompanied by perspective. And if it is so accompanied you may lack the principles of geometrical demonstration and the principles for the calculation of the forces and power of the muscles; or perhaps you may lack patience, so that you will not be diligent.

For these reasons, dissection was carried out in winter to slow the rate of decay and allow more time for appreciation of the parts under observation.

Perhaps as a result of these difficulties, Leonardo spent a considerable amount of time dissecting animals. A contributory factor to this would have been the acceptance of the commonality of much anatomy between the quadrupeds and humans. To that end, Leonardo clearly dissected several oxen. Most, if not all, of his cardiac anatomy is based upon the ox heart, as we at Papworth Hospital have recently confirmed by following his drawings and notes as a dissection manual. That he was accurate and precise we have clearly demonstrated. His dissection of animals, however, fulfilled another purpose. He was passionately interested in comparative anatomy, as well as in comparisons between the young and the old. This is exemplified in his dissection of a centenarian in Florence. Sitting with this old man in the Santa Maria Nuova, Leonardo witnessed his death and, intrigued by a passing 'so sweet', he rapidly made a dissection. He recorded the tortuosity of the blood vessels, and made note that he should compare these findings with both a child and the 'birds of the air and the

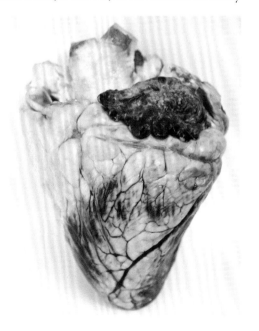

6 Ox heart

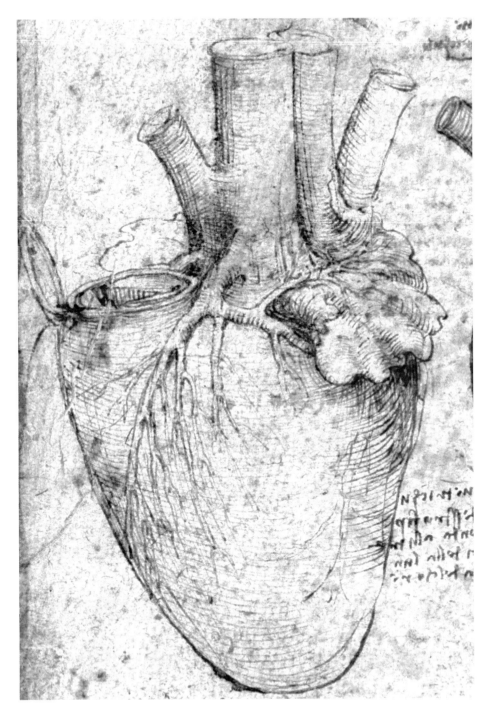

7 Leonardo da Vinci (Italy, 1452–1519), ox heart (detail of page showing mechanisms of the ventricles), c 1515–16, pen and ink on blue paper, 28 × 41 cm

beasts of the field'. He deduced that the tortuosity and thickening of blood vessels had led to a reduction of the blood reaching the affected heart, producing its malfunction and death. In this, he was surely the first to describe the atherosclerotic process, now the scourge of the Western world.

During his life, Leonardo had been engaged in large-scale irrigation schemes, both for peaceful purposes and for those of war, such as the attempts to divert the river Arno away from Pisa to defeat the Florentines' enemies in that city. It is possible that he was engaged in this with the political philosopher Niccolò Machiavelli (1469–1527). His observations of the effects of an expanding vortex of water from a narrow orifice into a wider space clearly influenced his ability to work out the closing mechanism of the aortic valve in the heart (more of which later), 455 years before anyone else.

Leonardo's most important contribution to cardiac anatomy came during the Vatican years from 1513, perhaps largely done at the Ospedale San Spirito, which lay within the Vatican walls. This early Renaissance building still stands, but is now enveloped by the modern hospital buildings. The only office now utilised within the confines of the original building is the cardiac rehabilitation department, a fitting vestige of its heritage. It is likely that Michelangelo also carried out dissection within its confines. In passing, it is worth stating that Michelangelo is known to have carried out a significant amount of dissection, and had associations with two or three professors of anatomy, chief among them being Realdo Colombo, a significant surgeon and anatomist. Since Michelangelo destroyed most of his drawings, little of his anatomy survives, and it is therefore difficult to ascertain his true competence in the field. There are some *écorchés*, but nothing to demonstrate an interest in, or knowledge of, organ systems. This is in stark contrast to Leonardo, who in this last period carried out a large amount of work on the functional anatomy of the thoracic and abdominal contents.

It is recorded by the chronicler of artists' lives Giorgio Vasari (1511–74), among others, that Leonardo had earlier developed a close working relationship with a bright and fast-rising star in the field of anatomy and medicine, Marcantonio della Torre (1481–1512). Their paths probably crossed when Leonardo was in Milan around 1503–5. Marcantonio was trained in Padua and established himself in Pavia. It is reported that Leonardo worked with him to produce an illustrated manuscript on anatomy. Sadly, no evidence of this exists today. It is quite likely that Leonardo would have watched and perhaps contributed to anatomical demonstrations by the young Marcantonio while at work within the University. Perhaps it was this association that fired Leonardo's interest in the true function of such organs as the heart, lungs, genito-urinary system and the magic of conception, pregnancy and birth, on all of which we have extensive notes and drawings in his hand.

In terms of the structure and function of the heart, much of what was held to be true at this time related to the work of Galen and Aristotle. Galen wrote extensively on the heart, but made several basic errors. Leonardo almost certainly came to him through the writings of Avicenna, said to have written 450 books, many on medicine, the most famous of which were *The Book of Healing* and *The Canon of Medicine*. Reputedly fully qualified by the precocious age of eighteen, and reported to have treated many needy patients without payment, his fame spread rapidly. He presented the doctrines of Galen, and through Galen, those of Hippocrates, but his interpretations distorted to varying degrees the original works of the masters. One stroke of genius would appear to be his disagreement with Galen over the interconnectivity of the heart and lungs. He was the first to challenge Galen's theory that the interventricular septum was perforated by multiple pores, 'for the substance of the heart is solid, and there exists neither a visible passage, as some would suppose, nor an invisible passage which will permit the flow of blood as Galen believed'. He suggested that blood 'crossed the lungs so that it can spread out in their substance and mix with the air and thence to the venous artery and from there to the ventricle'. This became known as the pulmonary transit. It is sad to think that either Leonardo was unaware of this gem, or ignored it. Perhaps he was familiar with it and used it, but that information is now lost, along with a substantial amount of his work. Galen continued to be a source of information into the sixteenth century. His attitudes towards the philosophy of life were similar to Leonardo's. Indeed, it is likely that the microcosm/macrocosm continuum that became a central tenet of Leonardo's thinking stemmed from the great man, who discussed these ideas in the second century AD.

Within Leonardo's cardiac research (for there is no more appropriate term for the work that he did), which was carried out from 1513 in Rome, he made a number of fundamental departures from perceived knowledge of the time, and several completely original observations. First, he recognised the heart as the centre and focus of the circulation. Galen had previously held that the liver was the originator of the vascular tree. Utilising a plant as analogy, with the roots representing the veins, and the stem and branches representing the arteries, Leonardo concluded that this was in fact the role of the heart. While he spoke of the almost spiritual role of the heart as the progenitor of life, his observations were grounded in his description of its functionality. For example, he wrote that the heart 'moves of itself and does not stop unless forever . . . marvellous instrument, invented by the supreme master' – even in this short passage, he manages to describe what we now refer to as the automaticity of the heart. In other words, he defines perhaps its most important property: that of self-generating, perpetual contraction. And despite this concept of invention by a supernatural being, he quickly gets back to earth in his description of the heart as a muscular pump with its own blood supply and a set of specific functions.

Galen's views on anatomy were gained largely through the dissection of monkeys and expressed in his *De Usu Partium*. In this work, he set out to enumerate all of the parts of the body and their function. Quoting Hippocrates, he said: 'All is in sympathy in the universality of parts and all the parts work together for the operation of each

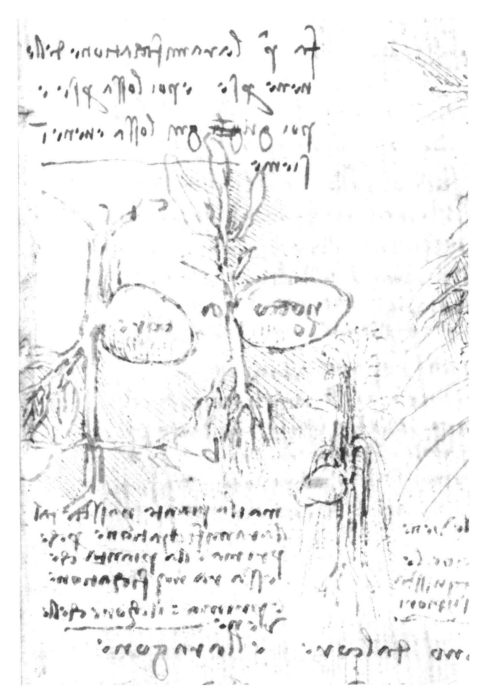

8 Leonardo da Vinci (Italy, 1452–1519), sketch of a plant sprouting from a seed that resembles a heart (detail of page showing arteries of the veins and thorax, the heart and its blood vessels), c 1508, pen and ink (two shades) over traces of black chalk, 19 × 13.3 cm

other' – words echoed by Leonardo, who states that 'nothing is superfluous and every-thing has a reason'. Galen completely integrated cardiac function with that of the lungs. While we know in modern physiology that these two vital organs are as they are to support, in unison, the life of the body, their functions are distinctly different. Whatever we may think of the heart today, it is first and foremost a muscle pump. If we set aside the advanced immunological function of the lung, its primary role is that of gas exchange, and it makes an important contribution to the regulation of the acidity of the blood by removing excess carbon dioxide produced by the body's metabolism. Galen held that by virtue of 'innate heat', the Vital Spirits of the body were manufactured by the mixing of air with the blood within the heart, and that the contraction of the heart led to the distribution of the Vital Spirits to all of the tissues. Hippocrates had described the heart as a three-chambered structure with the third chamber lying between the right and left ventricles. This myth was perpetuated by Avicenna and printed in Mondino's *Anathomia*. Galen, however, thought of the heart as a two-chambered structure. He did describe the valves within the heart, and entered

9 Depiction of a dissection, in Johannes de Ketham, *Fasciculus medicinae*, folio e2v, published by J and G de Gregoriis, Venice, 1495, woodcut

into discussion as to why the mitral valve possessed only two leaflets and the aortic valve three. These discussions can be found resonating in Leonardo's work nearly a millennium and a half later.

The method of teaching anatomy in the 1300s and 1400s is also relevant to our discussion because of Leonardo's association with Marcantonio della Torre and probably other professional anatomists. The kind of relationship between theory and practice that existed throughout this time is exemplified by the illustration of an anatomy lesson in Johannes de Ketham's *Fasciculus medicinae* (1495). Here is a wonderful example of the blind leading the blind! The professor of anatomy in his raised pulpit is reading from Mondino, while an assistant carries out the dissection as the students and possibly members of the public look on. The text of Mondino is a corruption of

Galen, and the facts being disseminated are unlikely to fit with what is being demonstrated, or worse, the dissection is being made to fit the descriptions within the text. It is probable that Leonardo would have witnessed scenes such as these in the medical schools, as well as being part of dissection within the artists' workshops and in his own studio, and perhaps even his lodgings ('passing the night-time in company with bodies flayed').

With the scene now set, it is possible to put into perspective the true genius of Leonardo and his accomplishments in the field of cardiac anatomy. It is telling that his last big effort in anatomy, during his Roman period from around 1513, is concentrated upon some of the great questions of physiology: namely, the structure and function of the heart and lungs, and the origin of life itself in his embryology studies. It was at this time that he demonstrated the full power of his questioning mind allied to great intellectual rigour. With this work, he made several important and truly original contributions.

Firstly, he correctly described the heart as a four-chambered structure, as well as the fact that both atria and then both ventricles contract in unison and sequentially. He described the moderator band, and rightly ascribed its function as a muscular bridge between the walls of the right ventricle, preventing over-distension.

Leonardo also continued Galen's discussion on why the aortic and pulmonary valves must have three leaflets to function correctly. In doing so, he put in place the thoughts that have today led to the concept of minimal energy surfaces in nature – the even sharing of load across a pressurised surface, such that every part of that surface

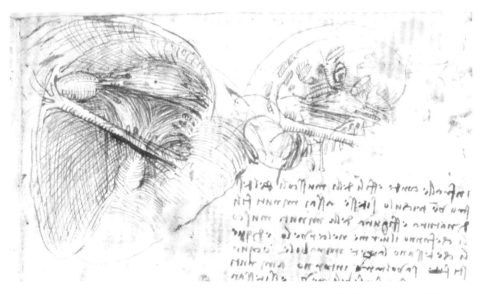

10 Leonardo da Vinci (Italy, 1452–1519), moderator band (detail of a page showing drawings and diagrams of ventricles), date unknown, pen and ink, 21.7 × 30.7 cm

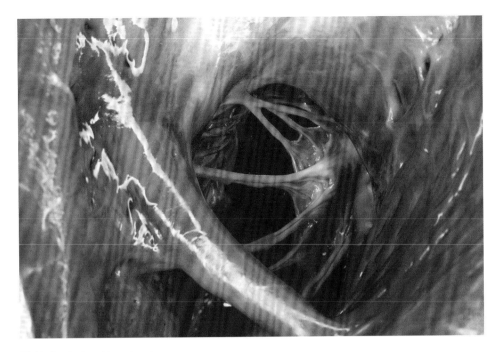

11 Moderator band in ox heart

bears the minimal load possible. This philosophy would lead to the deduction that the form of the load-bearing surface would mould itself in strength and form to the forces continually acting upon it. His use of geometry to explain the number of leaflets is beautifully revealed in a series of small drawings at the very top-right-hand corner of a page of notes, with a short written description. He shows how the greater depth of the leaflet in a four-leaflet configuration reduces the length of support from the other leaflets and increases the force on the leaflet when closing. Beneath these little drawings is a lifelike sketch of the valve fully open, through which one can vividly imagine the blood flowing with great force, emphasised by the fluttering of the leaflets.

In an exceptional piece of detective dissection, Leonardo also discovered and drew the bronchial arteries – small, paired arteries arising from the distal arch of the aorta, which supply blood to the walls of the major airways. He described with a great economy of words why those vessels are necessary as distinct from those that provide blood to the tissue of the lungs. These arteries are difficult to find, and few of even the brightest medical students, under direct instruction, are able to visualise them in the dissection room today. They play a vital role in lung transplantation, when the newly implanted lung is deprived of their nourishment, which can frequently lead to problems with bronchial healing. This is an example of how truthful observation and clever lateral thought can reveal crucial information.

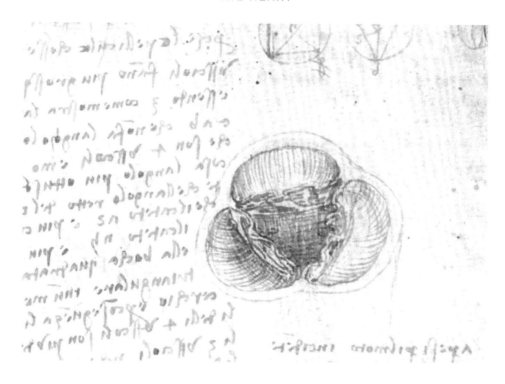

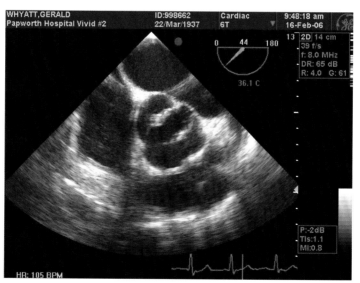

12 Leonardo da Vinci (Italy, 1452–1519), bicuspid valve (detail of page showing semilunar valves, and streaming and expulsion of blood), c 1506–8, pen and ink, 31 × 43.6 cm. **13** Human bicuspid aortic valve that has failed early, medical scan

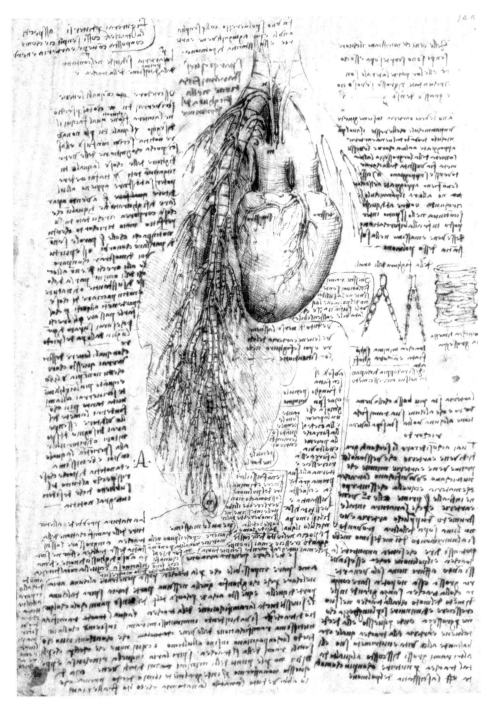

14 Leonardo da Vinci (Italy, 1452–1519), ox heart from the posterior view showing the dissected bronchial arteries and their relationship to the airways, c 1506–8, pen and ink on blue paper, 28.6 × 20.2 cm

While Leonardo's description of the purpose of cardiac activity remained disappointingly rooted in the Galenic tradition, his drawings of the interstices of the right ventricle and the tricuspid valve are quite breathtaking in their beauty and accuracy. He even describes plans for the making of a model of the tricuspid valve as a paper cut-out that, if made in his era, would enhance the teaching of the structure/function relationship of this valve. Indeed, this applies today just as well as when he wrote those notes.

The apogee of Leonardo's studies in anatomy, however, is his completely modern usage of hypothesis, experimental design and proof in the deciphering of the closure mechanism of the tri-leaflet aortic and pulmonary valves. In this piece of work, he seems to draw effortlessly upon all of his accumulated knowledge of engineering and

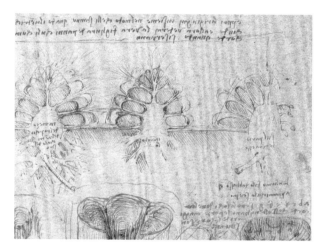

15 Leonardo da Vinci (Italy, 1452–1519), tricuspid valves (detail of stylised drawing of valves and ventricles), date unknown, pen and ink on blue paper, 28.3 × 20.7 cm. **16** Tricuspid valves in ox heart

hydrodynamic theory, and he combines this with lateral thinking to an extraordinary degree. As is the case in contemporary research, the most important element of such deciphering is to have adequate insight and education to ask the correct question of the problem in hand. Probably stemming from his acute powers of observation and his unshakeable belief that everything in nature is as it is for a reason, he found the complex shape of the aortic and pulmonary roots (the open-ended chamber above the valves that lead into the arteries of the aorta and pulmonary artery) an irresistible problem that required an answer.

It was not until 1968 that others seriously asked this question, when two engineers in Oxford, Bellhouse and Bellhouse, set about trying to find the answer, again through engineering and mathematical means. Essentially, the efficient and complete closure of these valves depends upon complex hydrodynamics, operating in the late phase of ventricular expulsion of blood. As noted, Leonardo had long been fascinated with the formation of vortices in nature, in the air as well as in water. His interest in these phenomena is most dramatically shown in his deluge drawings, c 1517–18, of a landscape flooded in a storm, and can be found in his writings and his engineering drawings. As stated earlier, a beautiful example is a drawing in which he depicts the erosion of the bank of a river by a vortex formed as a result of water flowing from a narrow opening into a wider channel. He applied this knowledge to the orifice of the valves in question and surmised that as the force of forward flow through the valve orifice ebbed away, the pressure exerted by the vortices behind the leaflets would overcome the force in front of the leaflets. He also deduced that not only would this force begin to close the valve, but it would also expand the semi-lunar shape of the valve leaflet and prevent it from being simply folded upon itself, as he argues would be the case if closure depended simply upon the reflux of the blood as the ventricle relaxed. Having made this hypothesis, he then uses rhetorical argument to discuss with himself why one theory is superior to another. When he had satisfied himself of the correct argument, he designed an experiment to test the hypothesis. In this, he blew a glass tube, with the sinuses (expansions in the wall above the seat of the valve) represented as a bulge, and probably inserted a handmade valve within the tube before the expanded area, mimicking true anatomy. In his own words, he then 'panicked grass seeds' within water in the tube to map the flow of blood as it passed across the valve and into the wider space beyond. From this emanated a series of drawings perfectly reproducing the actual vortices that occur in the natural state.

This elegant experiment remained unpublished, and was only rediscovered with the translation of his notes in the late nineteenth century, and lay dormant as a useful concept until 1968. Interestingly, since the state of knowledge remained behind that of Leonardo until that time, Dr Kenneth Keele, a Windsor GP and Leonardo scholar, commented that some of Leonardo's cardiac anatomy could not be confirmed at the time of the publication of his book on the cardiac work of Leonardo in 1952 because

17 Leonardo da Vinci (Italy, 1452–1519), *Map of the Arno*, date unknown, pen and ink and coloured washes, 23.6 × 41.6 cm

of the lack of experimental evidence. In the Bellhouses' first description of the phenomenon in their 1968 publication in the scientific journal *Nature*, they are able to quote only one previous description, that of Leonardo da Vinci.

No discussion of Leonardo's cardiac anatomy can be complete without some reference to the circulation of the blood. How could such a mind not have understood that the blood would pass round and round the body within a continuous circulation as described by the English physician William Harvey in his *Exercitatio anatomica de motu cordis*, published in 1628? Although Leonardo's discussions of the relationship between the heart and the lungs is Galenic, he does describe a circulation to the kidney, and the fact that urine production is proportional to the blood flowing to it, and he outlines a similar arrangement in the lungs. This does not equate, however, to a full circulation in the sense of blood entering through one channel and leaving through another. He recognised that a pulse palpable in any artery in the body was in time with the beat of the heart. He even attempted to calculate the cardiac output (the amount of blood that leaves the heart each minute). Surely he would have realised that the vast

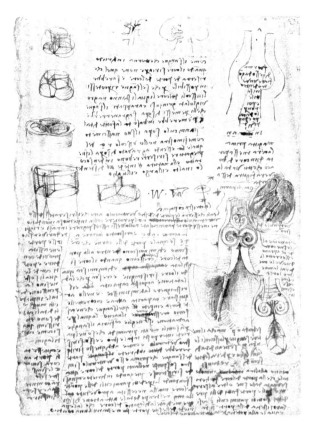

18 Leonardo da Vinci (Italy, 1452–1519), valve experiment (detail of studies of the heart), c 1516, pen and ink on blue paper, 28.3 × 20.7 cm

amounts that leave the heart must be circulating to account for the loss of volume. But all that we have is a description of ebb and flow associated with impetus – in other words, that the heart contracts, the blood is evacuated in part, and then after reaching the organ in question ebbs back to the heart and is revitalised before the process is repeated all over again. Perhaps he was so busy with other things that this was not a priority question. Perhaps he did get there and the information is simply lost to us or was withheld by the Church. We will probably never know the answer to that intriguing question.

Later, the graphic representations of the heart becomes more refined, but none compare with the consummate ease with which Leonardo blended structure to function through word and image in such an effortless and engaging way. The astonishing observations described by this giant of early scientific method were possible as a result of a sublime skill in graphic representation, and a mind unfettered by a rigorous Classical education, incubated in a natural environment, and enhanced by the wonderful and

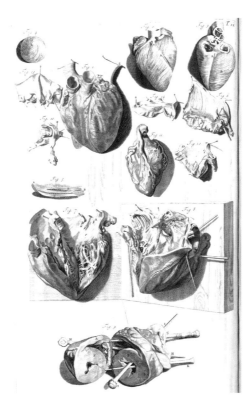

19 Drawings of the heart, in Govard Bidloo, *Anatomia humani corpois*, tab 22, published by Widow of J à Someren, Heirs of J à Dyk, Heirs & Widow of T Boom, Amsterdam, c 1685, engraving

varied world of Verrocchio's workshop. From the hill town of Vinci to the grandeur of the Medici court, Leonardo would have experienced a kaleidoscope of ideas, images and excitement.

Sadly, so much of his work is lost to us that we will probably never know the true expanse of Leonardo's vision. But despite this untold loss, he left us a treasure trove of material, to which we are still able to refer in order to make fresh observations using his insights. Indeed, when reading his notes, his combined talents of pure observation, clear deductive thinking and revealing drawings make one feel as though one is in direct communication with the master. If we devote time to looking and seeing through unblinkered eyes as Leonardo did himself, many more startling revelations will be likely to reveal themselves. We must be grateful for what remains of his work, due to the foresight of those who appreciated his genius in the early years after his death and those who kept it safe in the ensuing centuries.

Cosa bella mortal passa e non d'arte (Beauty in life perishes, not in art)

Leonardo da Vinci, *Codex Forster*

Francis Wells

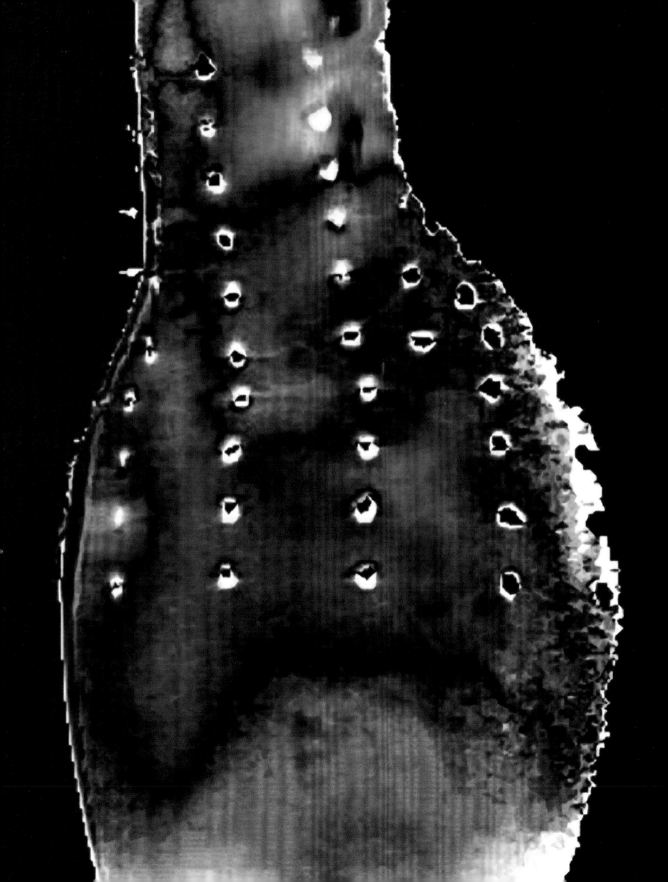

Francis Wells

Consultant Cardiothoracic Surgeon, Papworth Hospital, Cambridge, in conversation with Melissa Larner

MELISSA LARNER: When did you first understand that you had a vocation as a heart surgeon?

FRANCIS WELLS: When I was watching the nine o'clock news with my parents in 1967 and I saw Donald Ross, who latterly became a friend, being interviewed having done the first transplant in this country, and one of the first in the world. I think that fired my imagination, although I still had to complete my A-levels and go to medical school and qualify. [Laughs] And my initial suspicion that this was something I might like to do was confirmed very soon after starting medical school. Another student persuaded me to go with her to watch an open-heart operation at Charing Cross Hospital, in the very early days of this type of surgery, because she was a bit nervous about going on her own. I knew then that it was what I wanted to do. I planned the whole of the rest of my career around that goal.

ML: Why was it the heart in particular that drew you?

FW: It's the seat of everything, isn't it? If the heart isn't working, nothing else works. It's obviously *the* vital structure in the body, along with the brain. Life, to some extent, can remain even without a fully functioning brain, but you can't live without a heart, so there's no doubt about its immediacy, its vitality. Secondly, the challenge of operating on something that moves is exciting. There's nothing else in the body quite like that. And thirdly, there's the challenge that the surgery has to be as perfect, or as near perfect as you can make it; otherwise, you're found out very quickly, on completion of the surgery! So it really is a subject that captures and holds the imagination in a way that no other surgery does – for me anyway.

A model showing stress in the aorta due to sclerosis

ML: Can you describe that experience of working on something that's moving and beating?

FW: Yes. It is – even now – quite a moving experience, if I can use that unintended pun. Every time you embark on an operation, you know that you've got to get it right. So there's that focus, that immediacy, that feeling of in the moment when you're operating. Every time you step up to the plate to do one of these operations – whether you're not feeling so good, whether you're tired, or you've got some disaster in your life – you've still got to raise your game to that level of near perfection. That's something we've come to live with, that's bred into us over years of training. But nonetheless, it's there as a level of tension and anxiety, quietly beating under the surface. When you're learning, there are tremendous grey areas in trying to keep the heart safe while you're operating on it, and in doing so, keeping the patient safe. No one can really tell you with each individual heart what the limits are, where the boundaries are of safety and when you've strayed outside them. So it also brings yet another dimension to this kind of surgery, which is time. Now, we don't watch the clock, but we are aware of getting on with things. The decisions you make, you want to make in such a way that you're progressing through the operation one foot in front of the other with little pause, rather than taking sidesteps or backward steps, because you want to complete it all in a set time. So you're very engaged intellectually in the decision-making, right from the outset and all the way through.

And then, of course, something that really doesn't exist in quite the same way in other forms of surgery is the teamwork. It's a team effort. You can't do it without someone efficiently running the heart-lung machine, without a highly skilled anaesthetist getting the patient in the right state for you to operate, and helping them to recover at the end of it. You need the perfusionists to be experienced in order to run the pump safely. The nurses with whom you're working, and your assistant, need to be familiar with the environment, so that they don't waste time in not having the right equipment and facilities available. So all of this has to come together in a well-orchestrated effort, and I think in some ways, as a surgeon, you can liken yourself a little bit to a conductor with an orchestra. You have to bring it all together to make the performance right. There are many, many things going on in this operation, which again, I think is a challenge outside other forms of surgery. And at the centre of it all is Shylock's pound of flesh: this wonderful piece of meat, which is the most delicate and extraordinary structure in the body, and which has to beat from the first sign of life until the end of life.

ML: Do you think that other types of surgeons have the same kind of intimacy with their patients, or is there a special dimension, because the heart is this ultimate, life-giving organ and an emotional archetype in many people's minds?

FW: Without wishing to detract from anything my colleagues and other specialists do – because they'll have their own special form of relationship – I do think there is something special about this, because if you put yourself in the patient's place and think, 'This man's going to heave my chest open, stop my heart and fiddle about with it, and if it doesn't work, I've had it', it is a very special form of trust. And while we go to great lengths to explain to patients what's going on, and things are very standard in many ways in the routine parts of open-heart surgery, there's always that element of one human working on another human, both of whom are fallible. So that bond of trust is paramount: both the patient putting their trust in you, and you accepting that gift of trust and making the most of it to do the best for the patient, does produce a special bond. You see this when you talk to the patients and their families after you've finished operating. There's quite an outpouring of emotion when it's all over. I've done a lot of general surgery, I still do a lot of thoracic surgery, I operate on patients with lung cancer and so forth, which has its own emotional set of parameters, but cardiac surgery is something quite special.

ML: Are there any moments that stand out as particularly rewarding?

FW: Yes, there are operations that do stand out in your mind. I do a lot of valvular reconstructive work, and if you can keep someone's own tissues and then get them to function normally for the remaining lifetime of that patient, that's a very interesting and special challenge, especially at the extremes of what's doable. When you're in the middle of these operations, and the clock's running, you haven't got forever to fiddle about, and you have to make a series of decisions that take you down an operative path, some of which can be extremely challenging. And to come out the other end with the heart working beautifully and looking good, and the patient well, then yes, that's very dramatically rewarding. In an emergency situation, where you find yourself operating in the middle of the night and quite unexpectedly, and one minute you know nothing about the patient and twenty minutes later you're deep in their chest, you can find situations that are extremely difficult to deal with. And to get those patients through safely is immensely rewarding, and it can be very moving, too.

ML: What about the downside?

FW: It can be quite lonely, because a lot of this can go on at night. When you come out of the operating theatre, whether it's been a triumph or a disaster, you return to your office on your own to write up the notes, and reflect, and there's no one there to help you pick up the pieces if it hasn't gone well, or to celebrate with if it has. People tend not to see that. You can spend eight, ten, twelve hours operating on someone and then they die, and you come back, you tell the relatives, they're falling apart, you're feeling dreadful a) because you've failed, b) because of the emotion you've had to deal with, c) because you're exhausted. And whereas nurses have a wonderful backup system, you don't have that as a surgeon. There's nobody.

ML: And you have to break the news to the family at the same time as having to deal with your personal response to it.

FW: Oh yes, yes. And the tears do flow sometimes, too, on both sides of the fence. The only person who can really understand what you've just been through is a colleague who does a similar job. No one else can really understand what it's like: walking that tight rope, trying to close that stitch to a millimetre-perfect position, and dealing with tissues that are falling apart but upon which the life-blood of a patient relies.

ML: What would you say is the most common public misunderstanding about what you do?

FW: We've had a very good working relationship with the media here at Papworth, and I think there are some very responsible people who do a wonderful job, but also a lot of misrepresenting goes on. And the politicians have had their share, too, in undermining the trust between doctor and patient, through some well-celebrated disasters such as Bristol [where high mortality rates in paediatric heart operations led to a high-profile court case during the 1990s]. Some colleagues do abuse the situation and spend enormous amounts of time in private practice and don't devote themselves to the NHS as they should. These are very, very few, but they're then latched on to and set up as the mean for the rest of us, and I think that does undermine a lot of what we do. There's this assumption out there that we're all on the golf course and driving Bentleys, when in fact most of us are doing between seventy and a hundred hours a week, when well into our sixties, just trying to do our best for the patient. That irks. But it doesn't happen too often.

ML: If there was one scientific breakthrough that could change the face of what you do, what would it be?

FW: Something has already happened of that sort, and that's the introduction of angioplasty – dilating narrowed coronary arteries with the use of tiny tubes and balloons, and inserting little metal stents to hold the lumen open after it's been dilated. It's a wonderful treatment for patients, and it works. It's had quite a devastating effect on the amount of cardiac surgery in the coronary-revascularisation field. So I think we've passed through one of those eras. And if it were ever possible, it would make an enormous difference to open-heart surgery if we could have a drug that would halt the metabolism of all organ systems, so that they could be suspended for a significant period of time and then restarted. At the moment, we stop the heart beating, but we can't suspend metabolism completely. We stop the heart and we cool it right down to 12 degrees centigrade, so its metabolic rate slows down enormously, but it's still ticking over and the clock is still running and waste products are still being produced that will damage the heart if you leave it without a blood supply for too long. If there were a drug that could actually arrest all metabolism for a period of time, so you could basically do what you want, and take as long as you wanted over it to get everything absolutely perfect, that would dramatically alter a lot of what we do.

ML: Is there any chance of that happening?

FW: There's nothing on the horizon. Other things are happening, of course: preventative medicine has a huge role to play, both in terms of genetic predetermination of what may happen in the future, and the way people lead their lives. There's an ongoing development of equipment and devices and so forth, which continually makes our job more effective. There's a wonderful interplay between bioengineering and general mechanical engineering, the biological sciences and surgery, because we rely heavily upon various forms of devices, be they valve replacements, mechanical supports, pumps to run the circulation, suture material and so on. There's this tremendous interchange between all of these specialist areas, and every single one makes a little stride forwards here, a little stride there. It's like a sea front of knowledge and expertise slowly marching forwards – not by any means just by the surgeons, but by our colleagues in industry, without whom much of what we do today simply wouldn't be possible.

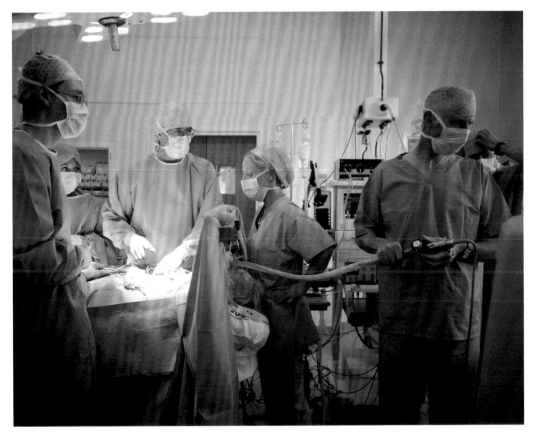

Ben Edwards, *The Moment of Truth* (Francis Wells operating at Papworth Hospital), 1998, photographic print, 106 × 138 cm

The Sacred Heart
Christian Symbolism

Emily Jo Sargent

If I have the Sacred Heart, what more do I want?

Marguerite Marie Alacoque

The heart, both that of Jesus and of the believer, has long been the site of devotion in Christianity. This is reflected in the variety and prevalence of images relating to the heart – particularly of the Sacred Heart of Jesus, which grew in popularity during the seventeenth century.

The focus on Jesus's heart developed out of the Old Testament. *Lev*, the Hebrew word for 'heart', or variations of it, can be found over 800 times in the Old Testament. It has been translated both as 'heart' and 'soul', which indicates the spiritual importance of the heart. These references rarely concern just the physical organ; the view passed down from the ancient world of the heart as the source of heat in the body and the location of the soul, memory and imagination had a pervasive influence on how the heart was understood, and what it represented. As such, the heart symbolised the person as a whole, while remaining the seat of their inner life, encompassing their thoughts and desires, as well as emotion and love; it was, literally, the centre of the body and soul.

In Christianity, the heart is primarily a place of mediation between God and man, the meeting point of the divine with humankind. God is love, and thus the heart, the enduring symbol of devotion and love, is the place where we encounter him. In the New Testament, the heart of Jesus becomes a metaphor not only for the divine on Earth, but also maintains a direct link between God and man. Jesus is the son of God and is a physical embodiment of the love of God, and yet 'loves with a human heart'. It is a reminder of his humanity, and a link with the divine. It also represents the love of man for God, and of God for man – a place to offer, and receive, love.

The meditation on the heart of Jesus is part of a wider veneration of the holy wounds, and the contemplation of the Passion. The side wound, from which blood and water flowed directly from Christ's interior, was seen as the gateway to his heart and his eternal love. Christ is literally open-hearted, inviting his congregation in. The wounds were considered a refuge, and a source of nourishment. Sometimes referred to as a 'well', the side wound poured forth the blood of the Eucharist and the water of baptism, and was the source of redemption and everlasting life. In John 7:37, 38 Christ says, 'If anyone thirsts, let him come to me: let him drink who believes in me', and we are told that 'rivers of living water shall flow from His breast.' The eminent Dominican mystic St Catherine of Siena (1347–80) experienced a vision of Christ who, proffering his wound to her, allowed her to do just that. Her confessor, Raymond of Capua describes it:

Bone IESV, fontes fluant,   Illis animam mundare
In cor nostrum toti ruant   A peccatis expiare
    Gratiarum riuuli.   Ecce gaudent Angeli.

Anton. Wiers fecit et excud.

1 Anton Wierix (Belgium, c 1555–1604), the Christ child as a fountain of life stands in the Sacred Heart with water streaming from His wounds onto the believers, c 1600, engraving with etching on paper, 7.9 × 5.6 cm

And putting his right hand on her virginal neck and drawing her towards the wound
in His own side, He whispered to her, 'Drink, daughter, the liquid from my side, and
it will fill your soul with such sweetness that its wonderful effects will be felt even by
the body for which my sake you despised.' And she, finding herself thus near to the
source of the fountain of life, put the lips of her body, but much more those of the
soul, over the most holy wound, and long and eagerly and abundantly drank that
indescribable and unfathomable liquid. Finally at a sign from the Lord, she detached
herself from the fountain, sated and yet at the same time still longing for more.

A reward for a life of self-sacrifice tending to the sick and needy, this moment shows
an intimate union between man and God, and heaven and earth. In a rare representation by Louis Cousin, the scene is presented as a heavenly vision, surrounded by angels.
Delicately painted on lapis lazuli, this object was probably produced as a private
devotional work for a rich patron.

2 Louis Cousin, called Luigi Gentile (Belgium, 1606–67), *Saint Catherine at the Wound of Christ*, c 1648,
oil on lapis lazuli, inset into panel, 24 × 27.3 cm

In some instances, we see Christ's wounds disembodied and bleeding as objects of devotion. In a late seventeenth-century German prayer sheet, for example, the side and shoulder wounds and the nail are presented independently for veneration. The shoulder wound, received from carrying the cross, is a rare inclusion, which did not inspire the significant devotional following that has been received by the side wound, with its direct access to Christ's heart. The accompanying prayer describes the side wound as 'honeysweet', and reads: 'O most healing wound of the heart of Jesus Christ, in your divine goodness I adore you and greet you. Blessed is the love which split you open, and blessed is the blood and water flowing from your side to wash our sins away.' These representations, like the text on St Catherine quoted above, possess an overt sensuality. The similarity to male and female genitalia reflects the relationship between the spiritual, divine love and a more physical yearning that is found repeatedly in Christian imagery. This was seen as a legitimate way of entering into a relationship with Christ, harnessing earth-bound passions in the service of an altogether holier love of Jesus.

In some instances, the side wound becomes a physical gateway to the heart. A painting by Bernardo Strozzi shows doubting St Thomas reaching into the side wound. It is a startling moment: Thomas penetrates Christ's body to access the love within. St Bonaventure (1221–74), an early advocate of the Sacred Heart, advised others to follow his example:

> Go then, go with all your heart to Jesus wounded, to Jesus crowned with thorns, to Jesus hanging on the cross and with the Blessed Apostle Thomas not only look at the signs of the nails in His Hands, not only put your hand in His Side, but enter wholly by the door in His Side into the very Heart of Jesus.

As the wounded heart of Jesus allows his love to pour forth, so a wounded Christian's heart allows the love of God to enter in. St Teresa of Avila, a sixteenth-century nun of exemplary piety, much given to visions – including one said to have lasted more than two years – encountered an angel 'in bodily form', who drove a lance into her heart. This action mirrored the creation of Christ's side wound. She writes:

> In his hands I saw a long golden spear and at the end of the iron tip I seemed to see a point of fire. With this he seemed to pierce my heart several times so that it penetrated to my entrails. When he drew it out, I thought he was drawing them out with it and he left me completely afire with a great love for God. The pain was so sharp that it made me utter several moans; and so excessive was the sweetness caused me by this intense pain that one can never wish to lose it.

While St Bonaventure and St Teresa were among those who made early references to the importance of the Sacred Heart of Jesus, it was a French nun, born in 1647, who

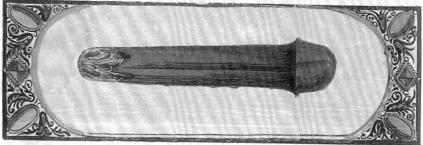

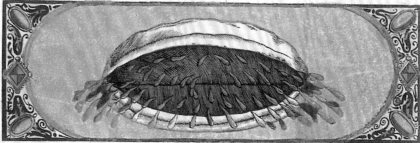

3 Prayer Sheet with the Wounds and the Nail, issued by JP Steudner, Augsberg, late seventeenth century

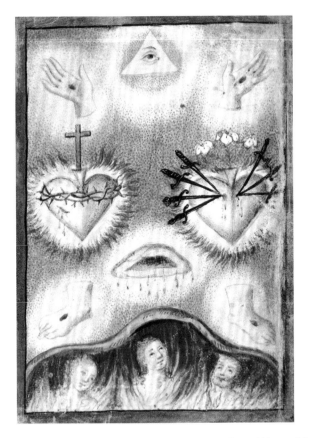

4 Anon, souls in purgatory looking up to the wounds of Christ and the Sacred Heart, eighteenth century, watercolour on vellum with gold border, 10.1 × 7.2 cm

was to instigate the formal devotion to the Sacred Heart. Marguerite Marie Alacoque showed an early predilection for prayer and piety, devoting herself to God and a life of chastity at the age of four. During her youth, she was obliged to care for her sick mother, who suffered from St Anthony's Fire, an unpleasant skin condition. Drawing strength from prayer and the contemplation of the Passion of Christ while dressing her mother's sores, Marguerite Marie entered into what was to become a life-long relationship with Christ. She developed a taste for humiliation and pain, and began to scourge herself to please Jesus. She wore hair shirts, iron chains on her arms and a cord tied tightly around her waist so that breathing and eating were painful. She slept in beds covered in splintered wood and broken china to feel at one with Jesus's suffering. When she was distracted from her path of piety by 'worldly pleasures', she experienced a vision of Jesus, his body bleeding, chastising her for her deviance. In 1671 she entered the Congregation of the Visitation at Paray-le-Monial, far away from the distractions of friends and family – Jesus had demanded that her heart should not be 'divided' by

5 Bernardo Strozzi (Italy, 1581–1644), *The Incredulity of St Thomas*, c 1620, oil on canvas, 89 × 98.2 cm

the love of others. The coat-of-arms of the convent showed the Sacred Heart of Jesus and the Immaculate Heart of Mary. She continued her practices of prolonged humiliation and self-punishment. She sought out the sick and, mirroring the work of St Catherine of Siena before her, brought herself to kiss their wounds and to suck the pus

6 Anon, *Saint Teresa of Avila*, eighteenth century, oil on canvas, 70 × 50 cm

from an abscess on a child's toe that was 'frightful to look upon'. Not long after entering the convent, she was rewarded for her dedication by continuing visions of Christ, and on 27 December 1673 he revealed to her his Sacred Heart for the first time. She describes it in her autobiography:

> Jesus Christ, my sweet master, showed himself to me, shining with glory. His five wounds were brilliant like five suns, and flames burst forth, on all sides from this sacred humanity, but especially from his adorable breast; and it opened and I beheld his most loving and most beloved Heart, which was the living fountain of his flames.

Jesus then performed an exchange of hearts with Marguerite Marie:

> He asked me for my heart. I begged him to take it; he did and placed it in his own divine heart. He let me see it there – a tiny atom being completely burned up in that fiery furnace. Then lifting it out – now a little heart shaped flame – he put it

back where he had found it. 'There my well beloved' I heard him saying 'that's a precious spark from its hottest flames. That will be your heart from now on, it will burn you up to your very last breath, its intense heat will never diminish – only blood letting will cool it slightly.'

It is interesting that the Jesus of Marguerite Marie's visions was familiar with contemporary medical practices – blood-letting was often used as a way to regulate the internal balance of the humours. The humoural system had been the dominant theory of how the body worked since ancient times and was the basis of medical treatment until the nineteenth century. Within this system, the heart was considered to be the source of heat within the body, and blood-letting was one method for alleviating any excess heat. This bears a significant resemblance to the burning hearts of Jesus and Marguerite Marie. However, while the Sacred Heart was gaining in popularity as a symbol in the Church, the humoural system, and the role of the heart within it, was being called into question in medicine. With the publication of William Harvey's *Exercitatio anatomica de motu cordis* in London (1628), the heart began to be defined by its mechanical function alone – as a pump.

Marguerite Marie continued to experience visions of Christ, in which he instructed her in various devotional practices, seemingly designed to appeal to her dedication to suffering. He demanded, for example, that she lie face down on the floor for an hour every Thursday night to console his heart. In a subsequent vision, he ordered her to organise a special festival dedicated to the worship of his Sacred Heart, calling her, 'the Beloved Disciple of the Sacred Heart, and the heiress of all Its treasures'. With the assistance and support of a young Jesuit priest, Claude de la Colombière, the dedication to the Sacred Heart of Jesus was born. Under his direction, Marguerite Marie wrote an account of the apparition, and in 1685 the first festival of the Sacred Heart took place. Her work complete, she died in 1690. Marguerite Marie Alacoque was canonised in 1920 by Pope Benedict XV.

The devotion to the Sacred Heart proved popular within the Catholic tradition, particularly with the Counter-Reformation. As knowledge of anatomy became more widespread, so the Catholic representations of the Sacred Heart became more realistic, more

IL CUOR DI GESÙ È IL TESORO DI TUTTE LE GRAZIE;
LA NOSTRA CONFIDENZA NE È LA CHIAVE.

7 Prayer card, 1898, coloured lithograph, 12 × 7 cm

8 Our Lady of the Seven Sorrows, eighteenth century, Italy, polychromed wood, 45 × 45 × 25 cm

visceral – an actual, physical heart shown disembodied, ready for veneration. It was not only the Sacred Heart of Jesus that was presented as an object for devotion, however; the Immaculate Heart of the Virgin Mary also gained in popularity in the seventeenth century. In works such as JPM Ruiz's *Los Sagrados Corazones, el Corazon de Maria*, we see her heart wreathed in roses, or with lilies growing from it – while in a companion work Jesus's Heart wears the ring of thorns and a cross. In an eighteenth-century Italian wooden devotional figure, her heart is externalised, pierced with seven swords. Each sword represents a 'sorrow', mirroring the prophesy of Simeon – 'Yea, a sword shall pierce through thine own soul (*lev*)' (Luke 2:35). The seven sorrows are: Simeon's prophesy, the flight into Egypt, the loss of the Holy Child in the temple, the meeting of Jesus and Mary on the Way of the Cross, the Crucifixion, the taking down of the body of Jesus from the cross, and his entombment. Originally, only five sorrows were adored – to match the five wounds of Jesus. This was later increased to seven, to match the seven joys (the Annunciation, the Nativity, the Adoration of the Magi, the Resurrected Christ's appearance to Mary, Christ's Ascension, Pentecost, and Mary's Assumption into heaven). While

Jesus's heart is a mediation between man and God, Mary's heart becomes a mediation between man and Jesus – her heart exerting power and influence over His.

At the same time as these devotional images and practices of the Sacred Heart were developing within the Catholic Church, the Protestant faith developed its own traditions relating to the heart. As the bread and wine of the Catholic Eucharist actually *became* the body and blood of Christ, so the Sacred Heart became an actual heart. As the Protestant Eucharist *represents* the body and blood, so the Protestant heart becomes a representation, a symbol divested of its corporeal function. That symbol is the one with which we are all familiar, one that has now become a secular symbol to represent love – without the messy business of blood and guts. During the seventeenth century, books of 'Emblems' were published, which featured this symbol over and over again in a series of situations intended as a guide to the various duties and sufferings of the good Christian heart. An effective method of disseminating devotional counsel to a wide congregation (particularly those who couldn't read), the Emblems were intended to serve as steps towards the process of spiritual enlightenment. Daniel Cramer, a Protestant theologian published his *Emblemata Sacra* in 1624, in which the unfortunate heart is shown being hit by a hammer by the disembodied arm of God, clothed in a cloud, accompanied by

9 Daniel Cramer, *Emblemata Sacra*, folios 17, 173, published by Lucae Jenissi, Frankfurt, 1624

the dedication: 'My heart is like a rock, the hammer softens me, I sustain the blow; why then, if only I might be better.' The heart continues its adventures, being chased by a dragon or becoming an eagle circling the wounded hands, feet and heart of Jesus. The Emblems range from the traditional heart symbolism of flaming hearts or hearts pierced with arrows, to the downright bizarre – a heart with wings shown riding on a snail as it crosses a bridge with the instruction 'Circumspecte'.

In the etchings of Anton Wierix, we see the young Christ preparing and protecting the believer's heart. As with the Emblems, the emphasis has shifted away from the physical heart of Jesus seen in the Catholic tradition to the heart of the Christian follower, a place to receive the love and grace of God. The heart becomes a site for personal contemplation and improvement, and the dwelling place for Christ.

It is easy to see the relationship between the symbolism and language of these representations of the heart, and the popular symbol of the valentine. We can compare the trials of the Christian heart shown in the work of Cramer and Wierix with those of the heart at the mercy of Venus in a late fifteenth-century print, Casper von Regensburg's *Venus und der Verliebte*, 1485. The heart again appears broken, burnt and pierced, this time in pursuit of love.

10 Anton Wierix (Belgium, 1552–1624), c 1600, engraving with etching on paper, 7.8 × 5.6 cm, **a** Christ cleans the believer's heart, assisted and venerated by angels. **b** Christ as Groom sleeping in the believer's heart, it is safe in wind and storm. **c** Christ preserves the believer's heart from false worldly decoration, captivity and pain

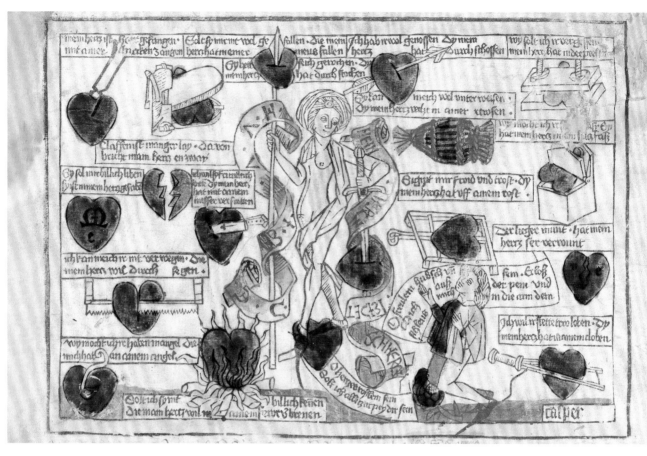

11 Casper von Regensburg (Germany, fl. c 1460–90), *Venus und der Verliebte*, c 1485, coloured woodcut, 25.7 × 36.5 cm

While the heart symbol used in Protestant publications has now become almost exclusively associated with the secular, the liturgy of the Sacred Heart remains in place in the Catholic Church – despite the wider cultural shift away from cardiocentricism. In the mid-twentieth century, amid charges of 'sentimentalism' (and at a time when heart surgery and ultimately heart transplantation were being developed) the Vatican was moved to re-examine the symbolic relevance of the Sacred Heart of Jesus. But with Pius XII's encyclical text, *Haurietis Aquas, On Devotion to the Sacred Heart* in 1956, the practice was validated. Here, the physical heart of Jesus was celebrated and reinforced as 'the natural sign and symbol of His boundless love for the human race' and 'as a sort of mystical ladder by which we mount to the embrace of "God our Saviour"'. The Sacred Heart of Jesus, and the secular heart alongside it, remains the ultimate and enduring signifier of love beyond any other. Despite the scientific understanding of the organ as a pump and the shift towards the brain as the location of self, it seems there is no clearer metaphor for love, divine or otherwise, than that provided by the heart.

INTERVIEW

Mike St Maur Sheil

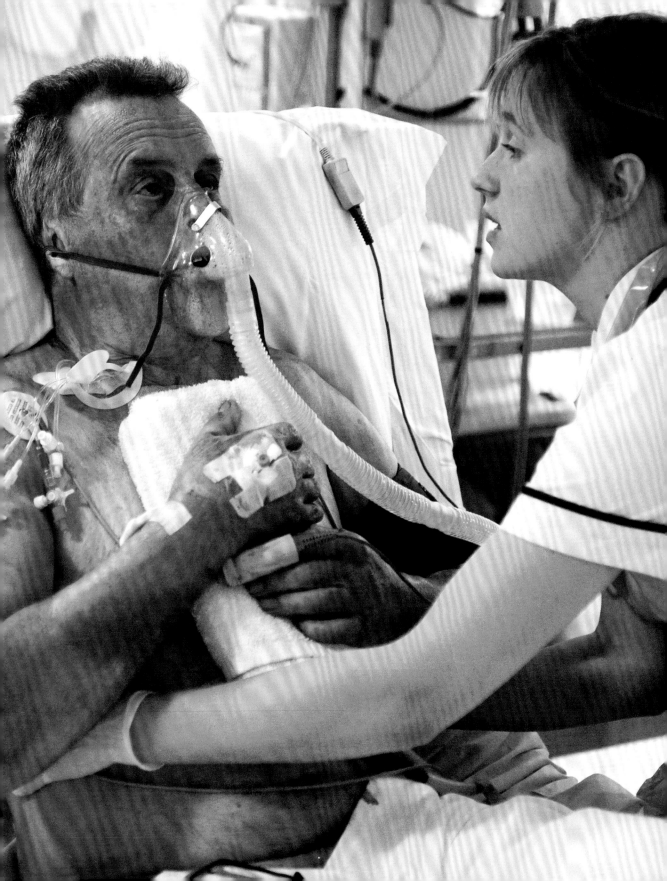

Mike St Maur Sheil

Photojournalist, in conversation with Melissa Larner

MELISSA LARNER: When did you first realise that you weren't well?

MIKE ST MAUR SHEIL: I was working for a construction company. I'm a photographer and I was doing some work on a building site. I was up some steel work, and I felt a little bit short of breath, and I had a headache – I just didn't feel quite right. That was the first intimation I had. I thought maybe I was just unfit, but then I had a recurrence over the next few days. Whenever I exerted myself, I'd feel a bit weird.

ML: And you hadn't had any chest pains before that?

MS: No, no chest pains at all. I'm normally very fit. I play sport and things like that. I went and saw my doctor and she said, 'It sounds like you've got angina.' And I thought, 'Yeah, cobblers. I don't have angina.' And she took a blood test and the cholesterol test came back absolutely normal. So I thought it must be stress. Anyway, she said, 'I've made an appointment for you to go and see a cardiologist.' And I went and saw a cardiologist who said, 'I want you back for some more tests. Don't do anything strenuous. Don't do any long drives.' In the meantime, I'm thinking, 'There's nothing wrong with me.' In fact, I was carrying on work as normal. I remember going up a 20-metre fire ladder! [Laughs] I did feel strange, but I carried on life as normal, and then I went into the hospital and they said, 'Hmm, interesting, yes, we'd like you to come in on Friday for a chest X-ray.' I went in on the Friday. I was asked a few questions and they wouldn't let me out. And on the Monday, I had an angiogram, at about ten thirty, eleven o'clock in the morning. About twelve o'clock the cardiologist came up and said, 'Just thought you might like to know what's happening,' and I said, 'Actually, I really do have to go

Physiotherapist Gemma Pitman helping a patient in the ITU, Bristol Royal Infirmary, 2005, digital photo by Mike St Maur Sheil

home.' He said, 'Well, you've got 99 percent occlusion. I've spoken to the Bristol, we're moving you there this afternoon and with luck they'll be operating on you tomorrow morning.' And I'm going, 'But I've got a job on Thursday!' So I was lucky. You hear these stories: 'Oh, did you hear about so-and-so? He was so fit. He died playing squash.' I was, according to the surgeon, within two to three weeks of that.

ML: And you didn't have a heart attack, which means your heart didn't actually sustain any damage.

MS: No. I'm told I have an extremely powerful heart, which is why I only had problems right at the last moment, when I had total blockage.

ML: But they couldn't sort it out with an angioplasty? There was too much blockage?

MS: With 99 percent occlusion, no. I had to have a double bypass.

ML: How did you feel when they told you that?

MS: At that point, I was sedated. You're just kind of like, 'Oh'. It's just one of those things. You don't really feel anything. I was very lucky, because I didn't even have time to think about it.

ML: Was there any point where you thought you were going to die?

MS: Um, someone came and asked me for permission to do the operation, and they said: 'There is a chance that things can go wrong, and in your case, there's an elevated risk: there's a 10 percent chance that the operation will not succeed.' In those days, when they came to clamp off the arteries, if they displaced any plaque from the inside of the artery, which is quite likely, and if that by any chance hits the occlusion, you're instantly going to have a heart attack. So the risk factor of the operation goes up. And I said something like, 'What if I don't have the operation?' and they said, 'Well in that case there's a 100 percent certainty of total failure.' So I said, 'Well, in that case, I don't have much choice, do I?' [Laughs] But I didn't have that horrible thing of, 'You've got to have an operation, but we'll have to wait until we can fit you in.' So I didn't have to worry about surviving until the operation. I was very lucky, in the sense that I was just told, 'Hey guy, you've got a problem, we're operating tomorrow. Sit back and enjoy the ride.' I didn't have any real trauma about facing surgery.

ML: How long did the operation take?

MS: Don't ask me! [Laughs] I think I went into surgery some time in the morning and, from what I'm told, I first became conscious again some time in the middle of the night, but really it was the next day before I came back, because they bring you back to consciousness very slowly to see if the heart is coping. And I had arrhythmia. So then, you're put back down again, and they get the heart to settle down, and they just bring you back up again very gently; it's a controlled resurfacing.

ML: And then it's a long time in hospital, presumably?

MS: No, it's not. Put it this way, the National Health Service is fantastic while the blues and twos are going on, but as soon as you're out of theatre, you're in a bed. And they want that bed! Five days is what they like to turn you round in. And believe you me, when you start eating their food, you can't wait! Five days is too long!

ML: And it's not very heart-friendly food is it? It's all stodge.

MS: Absolutely. I remember, the first time I managed to go for breakfast, I'm sitting there thinking, 'I've had a heart operation and I'm looking at sausages. Is this wise?'

ML: Did you feel quite emotional about your operation and grateful to the staff?

MS: Well, yes. I mean, there but for the grace of God . . . And here was a group of doctors and nurses who saved my life. It's as simple as that. I mean, they pretend they don't care, and I know they don't like to get emotionally involved with patients, so when you say, 'My god! You've saved my life', they say, 'Well, you know, I'm just doing my job.' But I saw my surgeon late one night, and I made a joke about, 'What are you doing here? You guys aren't supposed to care.' And he said, 'That chest has got my signature on it, so I care.' He was talking to me about heart-lung transplants, and he said that one of the most awesome sounds in the world is when you've taken the heart and the lungs out. If you click your fingers inside the cavity, it's like an echo inside a cathedral. And he said that someone comes in with a black bin liner, and you drop the heart and the lungs in there, and you've got about an hour to get it all wired up again. And he said another of the most awesome sounds in the world is to hear the heart monitor start to beep again. Because the heart is actually coming back to life.

ML: And how did your operation affect your family?

MS: My daughter didn't speak to me, really, for ten days.

ML: Why was that?

MS: I think she was so traumatised by what had happened. It was a real jolt – for everybody. And it was funny because I remember sitting at home and I made some comment to her, and she just suddenly came and sat on my lap and burst into tears. And that was the first real response I'd had from her since I got home. At the time, she would have been about eighteen. But she's a together lady. She's not someone who's given to emotional outbursts.

ML: How long did it take you to recover once you'd left hospital?

MS: A long time, actually. It was two months of hard work, and I found my chest very painful. Basically, somebody takes a Black & Decker, goes straight down the sternum, pulls it apart and then they wire it back together again. So, effectively, all your ribs are broken, and then anything you have to do with your arms is painful, because all the musculature is gone. I wanted to get back to absolutely 100 percent fitness. It was five months before I felt 100 percent able to carry on life as normal.

ML: That's with physiotherapy and exercise?

MS: Well, physiotherapy in those days – and I can't say what it's like now – but six years ago it was, quite frankly, bloody appalling. My local hospital, the John Radcliffe, was unable to offer me any remedial therapy whatsoever, apart from going to lectures on giving up smoking. You can't just go to a gym – they won't take you; you initially have to go to cardiac rehab, because you have to be monitored very carefully. It was very difficult to find physiotherapy. We finally did find a rehab unit, which was run by a really brilliant guy, and he was so brilliant he left. Because of short-term funding, he didn't know from one year to the next whether his unit would be in place the following year, so he finally said, 'Stuff this for a lark', and went off to teach at a university. But he had this really first-rate unit, which was regarded at the time by government as one of the centres of excellence, and yet the funding was chaotic. But people like the British Heart Foundation have these nurses who go out and do cardiac rehab, because really, if you have the operation and don't have rehab, the operation is wasted. It really is. You go through a programme, and after you finish the programme, then you can start going to a commercial gym, which is what I did, because I wanted to get back to being as fit and strong as I was before the operation, so that's why it took me a reasonable amount of time to do it. It's not as though I was going back to an office job: I needed to be able to go up steel works.

ML: And now you can do that?

MS: Yes. I've made a 100 percent recovery. In fact, I reckon I'm probably fitter for my age (sixty) now than I was at the time of the operation. But I consciously work at it. I'd actually decided I was going to give up cricket at that time, and part of my rehabilitation was to go back and play cricket again. So I did that. And I'm still playing cricket!

ML: So has this experience actually enhanced your life in some ways?

MS: Oh yes. Quite frankly every day is an extra day. I don't really give a damn about a lot of things that I used to care about, which were probably totally unnecessary, trivial things. And I've deliberately pulled back from being a fairly intensive commercial photographer. I now work on personal projects far more than I used to. So I'm not chasing a buck all the time.

ML: Is there more aftercare, six months, a year down the line? They check your cholesterol levels, don't they?

MS: Yes, that's a standard thing that your GP does, and I've been back to see my cardiologist a few times, but as far as I'm concerned – apart from certain things where my attitude has changed – my operation was a blip on the very distant horizon. I do think occasionally about it.

ML: You do feel some anxiety?

MS: I think every heart patient who's had an operation does a little. You do lie there in the middle of the night, sometimes, thinking . . .

ML: Is it going stop?

MS: Yes. And we're incredibly bad in this country. One of the things I've done, as a result of my experience, is I'm now what they call a 'Community Responder'. In rural areas where ambulances can't get to places within the government target times, a lot of local authorities are now doing this, and the idea is that you have people in communities who have had some basic cardiac and respiratory training, so if somebody has a problem, you can hopefully get there and help them. You're not there to replace the ambulance; you get there as quickly as possible so that you can start giving some preliminary treatment before the paramedics arrive. I've learnt how to do that, because of my experience, and it's a way of giving something back. It's something this country is very bad at. In most buildings, there's not even a CPR kit. This project first started in Seattle, where every single building that employs more than twenty-five people has to have a CPR kit. All public transport vehicles have CPR kits and the result is that of people who have a heart attack and receive CPR before the ambulance arrives, something like 50 percent survive. In this country, the figure is 4 percent. So, it's a very simple way to save lives.

ML: Tell me about the project with the hospital that you've been working on.

MS: I decided that as a way of saying thank you to the people at the Bristol Royal Infirmary, I'd like to take some photographs to document the work they do. So, in co-operation with the British Heart Foundation we went to a heart surgeon, Professor Gianni Angelini and said, 'Look, we'd like to take these pictures.' And Angelini said, 'Come along.' So I spent a week there, and with the permission of two patients, followed them through the whole process from coming in to going home.

And some of these pictures, people say, 'Urgh, gross!' But that's the reality of heart surgery. We live in an age where more and more people are going to have to have heart surgery. Everyone's talking about obesity levels. People should face up to the reality of what they're doing to themselves.

ML: But it's not always a result of one's life style, is it?

MS: In my case, apparently, I just produce very high levels of cholesterol naturally.

ML: Is it in your family?

MS: No, some people just do it. I eat an extremely healthy diet, I keep fit, I'm on high levels of statins, but even so I still have quite high cholesterol levels. It's going to get me one of these days. [Laughs] As they say, 'Life is a sexually transmitted terminal disease.' You know, it's one of those things. I've had my chance.

ML: What was it like, watching the surgery?

MS: It comes as a jolt, when you go in there and there they are – they're doing this incredibly intricate surgery, and they're talking about absolutely everything. They'll be discussing the music, the fact that the anaesthetist is wearing a funny hat, what happened in the England football match last night . . . But there's an incredible team working there.

  And the heart is an interesting piece of kit. What fascinated me during surgery was how durable it was. It's quite amazing. It just keeps on going. There's this thing pounding away, and they put a piece of gauze underneath it so when the surgeon wants to turn it over he just picks up the gauze and flops it – that turns the heart through 180 degrees, and it just continues to beat. It just carries on going. It's extraordinary.

ML: That probably helps you when you're having those midnight moments. At least you know how robust it is.

MS: Absolutely. It's got to be robust when you think about how many billions of times it beats during the average life. And there it is: there's a surgeon shoving needles in it and the thing still continues to tick away. Angelini and his team are now doing something that's becoming a standard operation around the world: they no longer stop the heart and put you onto a heart-lung machine; they actually operate on the beating heart. There are great technical advantages to doing that – there's much less trauma, because they're not having to introduce extraneous fluids into you, which they have to do with a heart-lung machine, and also you're less likely to experience brain damage, because if you use a heart-lung machine, micro bubbles of oxygen are introduced into the blood stream and these can get into the very fine blood vessels in the brain and cause tiny micro strokes within the brain. You don't get that with the off-pump surgery, where the heart continues to provide the blood throughout the body. But it is fascinating to watch the operating because Angelini has devised this chest clamp that they use to keep your ribs apart, which has a special foot on it, rather like the foot on a sewing machine, and he just puts that on the heart and moves it around and sews between the two feet. Angelini was an engineer; he actually trained as an engineer before he became a doctor. He's a plumber – a pretty super plumber! But because of his engineering background, he devised this special clamp to keep the heart still.

ML: That's interesting: two of the other surgeons I've spoken to have also had engineering backgrounds.

MS: Angelini is a man of real passion and pizzazz. And the Bristol is a centre of international excellence; they're pioneering operations and they're now developing a new operation: instead of going through the front of the chest they're going to go in under the ribs, because they can spread the ribs far more easily.

ML: Doesn't it make you feel strange, when you take these pictures, thinking that once, it was you lying on that table?

MS: Funnily enough, the bit that got me the most was not in theatre, but was in intensive care afterwards. It was just a little thing: here's this guy, I've met him a few times before, I've seen him being operated on, and here he is, and he's deeply unconscious, and he's wired up to all these machines that are beeping and ticking away, and I'm looking at him thinking, 'Are you going to live? Come on, come on, wake up, for Christ's sake!' There was a nurse, and she was washing him . . . and she was talking to him, and I . . . just found it incredibly moving, I really did, because I remember when I had my operation, how I was treated by one of the nurses. I started to vomit, and because you're wired together at the ribs, if you vomit, you run the risk of ripping the wire, and literally bursting apart. She realised what was happening, and . . . this nurse knelt down in front of me and . . . she held my chest together. She was hugging me and I was vomiting all over her. I was so upset, I was in tears. That to me, was the most moving thing, the way that nurse was so . . . tender. And that's what I remembered with this man, and that's the thing that I still find the most moving: the nursing. We all have this Florence Nightingale thing, and we're meant to fall in love with the nurses, but it's not that. It's their total humanity. I think the surgeons would be the first to admit that they've probably only met the patient for half an hour before the operation: you're totally anonymous, you're under theatre greens, you can't even see their face because it's screened off. Forget what you see on television, the face you see is the anaesthetist's; you don't see the face of the surgeon. This picture here of Angelini, that was a routine heart surgery, and yet, look at his face.

ML: He looks ravaged.

MS: Absolutely. Now that man, he's a very vibrant, charismatic character. He's incredibly fit. He was an Olympic-class athlete, but three hours in theatre and the guy is knackered.

ML: Did that guy in intensive care wake up?

MS: Yes, he did. The Bristol BRI has a very high survival rate, although they're pushing the limits the whole time. They're always trying to find better ways of doing things.

ML: I spoke to a surgeon yesterday who said what he would really hope for is to eliminate surgery altogether. He was also saying that teamwork was also a key thing, which is what you must have seen.

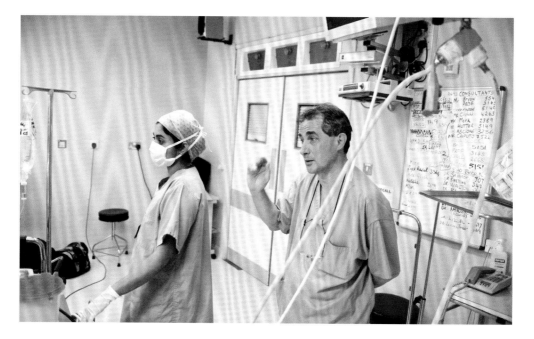

Professor Gianni Angelini after completing a four-hour 'off-pump' heart bypass, 2005, digital photo by Mike St Maur Sheil

MS: Absolutely, communication is everything. It's a very, very intense group of people working together. They know each other very well. The surgeon may be the star, but he doesn't function without the humblest theatre nurse. Every single one has a job. For example, they have a transparent plastic sheet with pockets in it, and there's a nurse who's responsible for ensuring that every single thing that goes onto the table comes off the table and gets put into this bag. So they literally count every single pin.

ML: That whole idea of surgeon as God seems to be changing.

MS: But they've still got a problem. They're still taking risks. They have to, because we all require more and more sophisticated treatments, and as they get better, they can do things that they never used to be able to do. They're now operating on people whom, six or seven years ago, they wouldn't have operated on, because they'd have regarded them as basically a lost cause. And the surgeons are taking risks, but of course, they're judged on how many patients they fail to save. And if you've got a pioneering surgeon, he's trying to save people whom ten years ago would almost certainly have died at home, and if they die on his table that's a black mark against him. But he's actually trying to save lives and it's a real dilemma for surgeons. I think they're very courageous to take the risks they sometimes do. The top guys are always trying to extend the boundaries, which is what we expect from them. They're trying to do their job better and better all the time.

The Emotional Heart
Mind, Body and Soul

Fay Bound Alberti

Introduction

The heart has long been regarded as an emotional organ. Descriptions of emotional experiences as 'heart-breaking' or 'heart-warming' (and of compassionless individuals as 'hard-hearted' or 'heartless') are more than turns of phrase. Rather, they reveal the heart's cultural legacy as the physical and symbolic space in the body where emotions are felt and experienced.

In the modern age, however, it is the brain that is reified as the site of our memories, our emotions, our 'selves', the organ associated with the presence of life or, indeed, its absence. To be 'brain dead' is now the official definition of death in the Western world, where once it was the cessation of the heart beat. The heart, by contrast, has been relegated to the role of a pump that distributes blood around the body. And yet it continues to carry significant cultural importance. It endures as a symbol of love and authenticity. We feel emotions in our chest, talk about our hearts being 'swollen' with pride or 'broken' by grief, and of acting 'from the heart'. All these terms convey a shared, unspoken assumption that the heart possesses some form of pre-cognitive intelligence: the heart *knows*.

But what is it that the heart knows? And how? To understand its enduring emotional status in today's neurocentric world, we need to explore the heart's origins as both an anatomical object and a cultural symbol. We need to trace the medical history of the heart from the Classical period to the present day, exploring how emotions were once believed to be products of the soul, linked to the body via the heart, as the single most important organ. From the second to the seventeenth centuries AD, a humoural theory of the body predominated in medical and scientific theory. Humouralism provided concrete links between the heart and such experiences as rage and sorrow, and created a language of the heart – that 'bursts' with joy or 'chills' in fear – which remains in common usage. As we shall see, after the seventeenth century the heart was intellectualised, at least in medico-scientific terms, and the brain prioritised as the site of intellect, feeling and cognition. Yet the emotional heart beat on.

The Emotional Heart of Antiquity

Since antiquity, the heart has played a crucial role in understandings of the body and the mind. In the fourth century BC, the Greek philosopher Aristotle described the heart as the centre of life, sensibility and feeling. 'Passions' like anger were felt as a 'seething heat in the region of the heart' because that organ was seen as the origin of life, sensibility and feeling. This interpretation was more than metaphorical. The heart was conceived as a three-chambered organ that was literally and figuratively the centre of the body and the mind. All other organs, including the brain, existed purely to cool the heart.

Unlike modern interpretations, ancient theories did not recognise distinctions between 'mind' and 'body'. There was no notion of a nervous system paralleling the heart and arteries, and the body was not imagined as a purely material structure. Belief in the soul helped to explain how the body was structured and functioned. Under the instruction of the soul, the material components of the body were allotted certain tasks. The heart was assigned the role of 'beginning' or 'origin', as a source of heat, and as the central organ of sense. As the source of innate heat, the heart mediated between the senses, the passions, the soul and the brain.

This link between the emotions and the heart was challenged by early anatomical investigations. In the medical school of Alexandria c 300 BC, experiments by Erasistratus and Herophilus (335–280 BC) into the nervous system suggested that it was the brain rather than the heart that was at the centre of sensation and cognition. Erasistratus elaborated his view of the *pneuma*, which endowed an individual with life, perception and motion. He believed that this substance was the cause of the heartbeat and the source of the body's innate heat. A distinction was made between the vascular system, the heart and arteries, and the abdominal organs – where life was controlled by the Vital Spirits – and the nervous system (where another substance, Animal Spirits, controlled motion, sensation and the senses). This interpretation of the 'spirits' that coursed through the body remains with us linguistically: we are 'high' or 'low' spirited, depending on our mood.

According to anatomical principles, the status of the heart had diminished here in comparison to its role for Aristotle. And yet it continued to be significant as the site where emotions were most universally *felt*. As Hippocrates saw it, this was largely because the organ was physically linked to the rest of the body. Many of Hippocrates's

1 Dumont, *Portrait of Galen*, c 1810, lithograph on paper, 33 × 23.7 cm

theories were set down by Galen, whose own interpretations of the body and mind in health and disease gained unparalleled longevity. In the second century AD, Galen established a system of medical practice in which the heart and the passions were anatomically and spiritually situated as mediators between mind and body. This system would be unchallenged until at least the seventeenth century. Indeed, recent research into therapeutics suggests that it was not until the nineteenth century that medical practice ceased being regulated by traditional principles of harmony and the humours.

The Galenic Model: Heart and Humours

In the 2,000 years after Hippocrates, 'Galenism' emerged as a set of connected principles that dominated psychological and physiological theories throughout Europe. Galen set Hippocratic medicine within a broader anatomical-physiological framework, providing material and spiritual explanations for emotions and for the links between mental and bodily experiences. He also attempted to explain why it was that individuals apparently experienced a variety of emotional tendencies, depending on

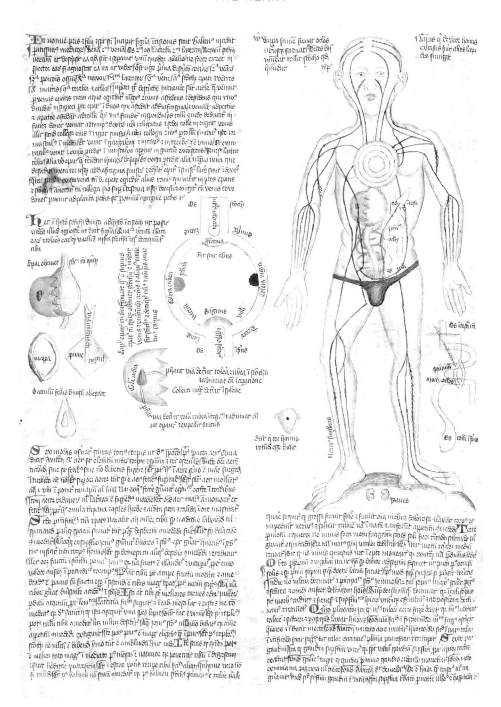

2 Organ Man, figures of arteries, organs, etc, in S Johannis, *The Apocalypse*, folio 36 verso, c 1420–30, ink and watercolour

their nationality, gender or age. Before we consider these in any detail, however, let us consider the anatomy of the body according to Galen.

Like Aristotle, Galen viewed emotions in the context of a body and mind that were linked to one another, as well as being influenced and guided by the soul. Galen's conception of the body was tripartite, its three parts mirroring the structure of the universe. The domains of heaven, sky and earth were therefore paralleled with the three main parts of the human body: the head (used for reason), the breast (associated with the heart) and the lower body (important for nourishment and procreation). Within this system, the body was physically composed of the 'greater world' of the elements: fire, air, water and earth. Each of these elements had a certain quality with which it was associated: heat, coldness, wetness and dryness. Fire was hot and dry; air was hot and moist; water was cold and moist; and earth was hot and dry. These characteristics provided the blueprint for each physical constitution in the form of 'humours', which were believed to course through the body. These included blood, which was hot and moist like air; choler (or yellow bile), which was hot and dry like fire; phlegm, which was cold and moist like water; and melancholy (or black bile), which was cold and dry, like earth. A recent illustration of the proportional balance of the humours vividly illustrates the four humours along with the four elements, the four seasons and the four ages of man.

The emotional life of individual men and women was believed to be influenced by the quantity and distribution of these humours, as expressed in language that we recognise today. A high level of yellow bile, for instance, made men and women subject to anger (choler) and black bile to sadness (melancholy), while an excess of blood or phlegm made one sanguine (and prone to love-sickness), or phlegmatic. An individual's 'dominant' humour depended on a variety of internal and external factors, including heredity, age, sex, and what contemporaries called the six 'non-naturals': air; food and drink; exercise and rest; sleep and waking; evacuation and repletion; and passions of the soul. Humours were therefore environmental as well as constitutional. As the seventeenth-century English philosopher Thomas Hobbes (1588–1679) put it, their proportions 'proceedeth partly from the different constitution of the body, and partly from different Education'. Yet humours, and emotional states, were inherently gendered.

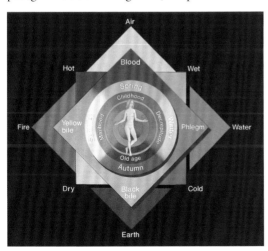

3 Lois Hague, *The Four Elements, Four Qualities, Four Humours, Four Seasons and Four Ages of Man*, 1991, airbrush on artist's board with acetate overlays, 35.8 × 48.1 cm

Gendering Emotion

Each age and sex was accorded specific emotional characteristics. In *The Passions of the Minde* (1604), written by the Catholic priest and religious controversialist Thomas Wright (c 1561–1623), male and female emotional tendencies were clearly set out. As a general rule, young men were viewed as 'hot, incontinent and bold', while women were 'envious, proud and inconstant', and these differences were a result of their bodily make-up. The gendering of emotional characteristics under humouralism reflected the physical differences associated with men and women. Women tended towards a phlegmatic or cold and moist disposition, since their bodies were fleshier, softer and weaker than those of men, their hair longer, their faces paler, and their skin more moist. The greater passivity of women also made them more subject to such emotional extremes as hysteria. By contrast, men had leaner bodies and drier complexions, and these were associated with qualities of courage and anger. Age was influential along with gender – men in particular displayed a variety of emotional temperaments during the course of their lives. While 'younge men' tended to be 'arrogant, prowde, prodigall [and] incontinent', old men were 'subject to sadnesse caused by their coldness of blood'. Expressions of emotion were also gendered. Because women had more moistness in their bodies they were inevitably more prone to tears, as well as to sudden, irrational rages, since women's flesh 'is loose, soft and tender, so that the choler being kindled, presently speeds all the body over, and causeth a sudden boyling of the blood about the heart'.

As illustrated by Wright, the physiology of passion was rooted in the heart. While the brain was increasingly identified as the seat of reason, the heart was crucial to Galenic theories of the emotions, just as it had been for Aristotle. Although Galen rejected Aristotle's claim that the heart was the origin of the nerves (prioritising the liver as this was the organ where humours were produced), he described the heart as 'the hearth and the source of the innate heat that vitalizes the animal'. Noting its unusual physical properties – 'a hard flesh, not easily injured' – Galen claimed that the expansion and contraction of the heart demonstrated its intelligence: the heart moved independently to attract what was desired and to repel what was unwanted. This contraction or expansion (what we term systole and diastole), took place with a cooling or a heating of the humours around the heart. In the case of anger, the blood became hot. Compare Pierre de la Primaudaye's conduct-book description of rage, for instance (1577), with the preaching of the English divine, John Downame (1609), each of which defined anger in terms of a 'boiling' of the blood around the heart:

> For first of all when the heart is offended, the bloud boyleth round about it, and the heart is puffed up: whereupon followeth a continuall panting and trembling of the heart and breast. [Primandaye]

[Anger is] an affection, whereby the bloud about the heart being heated, by the apprehension of some injury offered to a man's self or his friends, and that in turn, or in his opinion onely, the appetite is stirred up to take revenge. [Downame]

In Hot Blood: The Heated Heart

Ancient belief in the heart as an agent of heating was compatible with emotion beliefs that linked the mind, the body and the soul. Because Renaissance physicians regarded the soul and the body as indivisible, the soul was necessarily involved in the production of emotions. This gave emotions a moral or ethical significance. Galen wrote of the heart as enlarging and contracting to attract or repel external forces. What the mind interpreted as good (and worthy of attraction) or evil (and repellent) was determined by the soul. As Wright explained, the emotions were 'operations of the soule, bordering upon reason and sense, prosecuting some good thing, or flying some ill thing [and] causing there withall some alteration in the body'.

When the soul signalled the spirits to move towards or avoid an object, the heart's role in the process was crucial. In order to implement the desires of the soul, the heart required humours: melancholy blood for pain and sadness, blood and choler for anger. The humours were concentrated in the heart before being sent around the body to perform the operations of the soul. The agitation of the spirits in joy, and the free flow of the blood throughout the body, was in direct contrast to the physiological experience of fear, when the blood retreated and the soul shrank back from a perceived threat. This process explained the physical signs of emotion and the loss of bodily control that seemed to accompany extreme emotions. It also echoes modern, 'hydraulic' visions of the emotions as products that build up under pressure and threaten to burst forth if not released gradually, and in timely fashion. In Downame's *Treatise of Anger*, for instance, the author observed that anger made 'the haire to stand on end . . . The eyes to stare and candle . . . The teeth to gnash like a furious Bore'. At the same time, the tongue began to:

> stammer, as being not able to express the rage of the hart. The bloud ready to burst out of the vaines, as though it were affraide to stay in so furious a body. The brest to swell, as being not large enough to contain their anger, and therefore seeketh to ease it selfe, by sending out hot-breathing sighes.

If anger caused the blood to boil around the heart, the reverse physiological process was associated with fear, grief and sorrow. As Walter Charleton (1619–1707), physician to both Charles I and Charles II, and President of the Royal College of Physicians, explained in his 1674 treatise on *The Natural History of the Passions*, these 'negative' emotions caused the soul to contract. Thus in grief and sorrow, 'the Animal Spirits' were:

recalled inward, but slowly and without violence: so that the *blood* being by degrees destitute of a sufficient influx of them, is transmitted with too slow a motion. Whence the pulse is rendered *little, slow, rare* and *weak*, and there is felt about the heart a certain oppressive *strictness* as if the orifices of it were drawn together, with a manifest *chilness* congealing the blood and communicating itself to the rest of the body.

Such 'dejecting symptoms' had long-term detrimental effects on the body. They created dangerous levels of coldness that could, according to Charleton, 'incrassate the blood by refrigeration, and by that reason immoderately constringe the heart, cause the lamp of life to burn weakly and dimly'. Perhaps unsurprisingly, melancholia, even death, could result. As these examples illustrate, early descriptions of the blood 'boiling' or 'cooling' around the heart were not only metaphors, but described a perceived physiological process. We use this heart-based language of emotions today, though its historical associations have long been forgotten.

With the anatomical revival of the Renaissance, many of Galen's theories were cast into doubt. Yet Galenic explanations of the emotions as products of the body and the mind; as subject to distinctions of class, age and gender; and as linked to the heart, remained intact. Throughout Europe from the seventeenth century, the heart was subject to investigation and experimentation, particularly in relation to the brain, the cardiovascular and the circulatory systems. In part, this can be attributed to the rise of anatomical dissection (which challenged existing theories of the mind and the body), but also to philosophical and scientific digressions about the relationship between humans, God and animals, and the statuses of the mind, body and soul. How far did such discoveries as the circulation influence the decline of the emotional heart in science? Is the origin of the unemotional heart in science to be found in its reconceptualisation as an organ responsible for the movement of the blood, rather than the origin of heating, attraction and repulsion?

Circulating Theories: Anatomising the Heart

In the Western medical tradition, theories of scientific advance focus on the post-Renaissance revival of anatomy. Yet anatomical investigation was conducted much earlier in medieval Islam, where the discrepancies between the work of Aristotle and Galen gave rise to detailed consideration of the heart's role. Contemporary anatomical drawings, aiming at simplicity rather than naturalism, charted the flow of the arterial system. The Persian physician, philosopher and scientist Avicenna, was the author of

4 Anatomical illustration of a human figure showing arteries and viscera, from a manuscript attributed to Shikastah-Nastaliq, eighteenth century, Persia, drawing, 50 × 15 cm

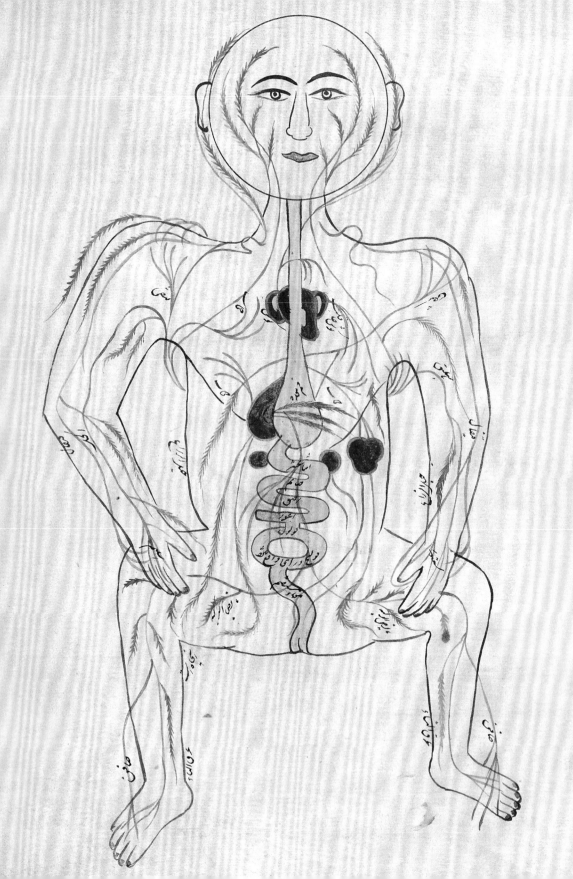

The *Canon of Medicine* (1030), a work that became a standard teaching aid throughout Western Europe for seven centuries. Here, Avicenna perceived the function of the heart as determinant of general bodily health, and argued that the health of the heart could be impacted upon by the emotions.

Avicenna's ideas on the heart were taken up from 1236 AD by Ibn al-Nafis, who challenged Galen's idea of two separate circulatory systems and provided the first published account of pulmonary circulation (see Chapter 1). Anatomists in Renaissance Europe developed these ideas further in a series of experiments into the structure and function of the veins, the arteries and the movement of the blood. Realdo Colombo, in

5 Title page of Realdo Colombo, *De re anatomica*, XV libri, published by N Beuilacqua, Venice, 1559, engraving

his *De re anatomica* (1559), included a description of the movement of the blood through the lungs and back to the heart. The Italian physician Andrea Cesalpino (1519–1603) supported the principle of the pulmonary circulation, and was one of the first to use the term 'circulation' to describe the movement of the blood. And in 1603 Hieronymus Fabricius wrote *De venarum ostiolis*, which described the venous valves and their role in ensuring efficient distribution of the blood around the body.

Of course, not all anatomists were convinced of the existence of pulmonary circulation. In *De humanis corporis fabrica* (1543), the Flemish anatomist Andreas Vesalius described the means by which 'the blood sweats from the right into the left ventricle through passages which escape the human vision', before discussing the two circulatory systems separately. By the time that the English physician William Harvey published his findings on the heart and the blood, there were clearly existing debates over the possibility of blood circulation. Several of these, such as *De re anatomica*, Harvey cited in his own work.

As described in Chapter 2, Harvey, in his famous *Exercitatio anatomica de motu cordis*, demonstrated the reality of pulmonary transit, as well as connections between the arteries and the veins. *De motu cordis* is traditionally understood to have transformed contemporary ideas of the body. Yet Harvey was no radical. Though he conceptualised the body in mechanistic terms, he was essentially Aristotelian, retaining an understanding of the body as moved by vital forces. The blood that he described was not the same substance that we understand today, but 'vaporous, full of spirit'. Moreover, it was heated and cooled according to traditional principles. Thus Harvey's account of the heart was compatible with Galenic interpretations of the emotional heart (with links to the spirits, the soul, and heat and as the body's symbolic centre):

> So in all likelihood it comes to pass in the body, that all the parts are nourished, cherished, and quickened with blood, which is warm, perfect, vaporous, full of spirit, and, that I may so say, alimentative: in the parts the blood is refrigerated, coagulated, and made as it were barren, from thence it returns to the heart, as to the fountain or dwelling-house of the body, to recover its perfection, and there again by naturall heat, powerfull and vehement, it is melted, and is dispens'd again through the body from thence.

Harvey's heart remained the 'fountain or dwelling-house of the body'. In analogies intended to echo the monarch's unrivalled position in the country (and ironically, soon before Charles I's bloody clash with Parliament), Harvey described the heart as 'the beginning of life, the Sun of the Microcosm. The King ruled over his subjects, and the heart over the body, just as 'the Sun deserves to be call'd the heart of the world'.

The Unemotional Heart: The Mechanistic Pump of Science

Rather more threatening to the emotional status of the heart in medico-scientific theory was the work of mechanistic philosophers like René Descartes (1596–1650), who adopted Harveian principles in their emergent philosophies. In *Les passions de l'âme* (The Passions of the Soul, 1649), Descartes drew on contemporary materialist thinking, and examined the physiological implications of the new mechanical philosophy. It is with Descartes and his potentially radical split of mind and body, emotion and reason, that we find historical evidence of the emergent unemotional heart of science.

Descartes viewed mind and matter as incommensurable, rather than united, as they had been in philosophical approaches dating back to the Classical world. In Descartes's writings, the material body was corpuscular and mechanistic; the mind or soul immortal. Reason was the principle that identified the operation of the soul in

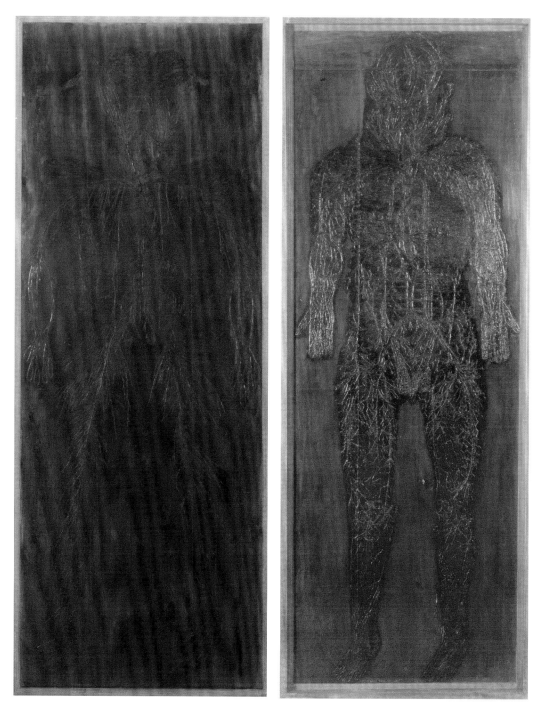

6 William Harvey's anatomical tables, venal system (female, left) and arterial system (female, right),
c 1640, cedar wood with varnished human veins and arteries, 200 × 75 cm

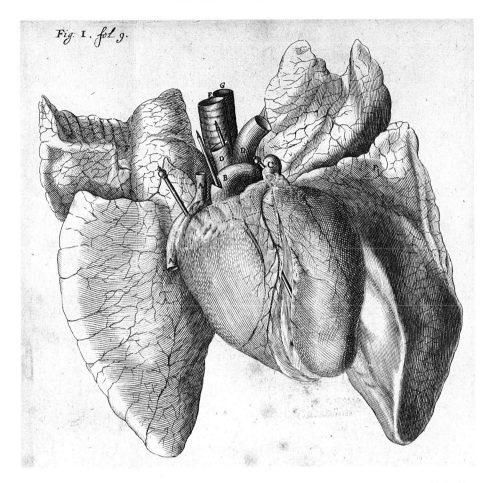

7 Drawing of a human heart, all layers lowered, in René Descartes, *De Homine*, folio 9, fig I, published by F Moyard and P Leffen, Leiden, 1662, engraving

humans, and the soul was given a concrete site of interaction – in the pineal gland of the brain. At its most simplified, this philosophical theory relegated the heart to a mechanistic pumping device that moved the blood around the body. The extent of Descartes's materialism is debatable. He did not deny the existence of the soul, and was in many ways traditional in his conception of the body's workings. He believed, for instance, that the passions mediated the mind and the body. He maintained that they were linked to the soul (and associated with the perception of good or evil), and that they had distinct physical effects. But the heart was no longer the site of emotions, or involved in physiological emotional responses.

In focusing on the doctrine of the reflex – which helped explain the machine-like functioning of the nervous system and the body – Descartes's writings potentially removed the soul from its position as a causal factor in bodily changes. At the same

8 Jurrianen Pool (Dutch, 1666–1745), *Cornelius Boekelman, Johannes Six, Twee overlieden van het Amersterdamse chirurgijnsgilde*, 1699, oil on canvas, 74 × 117 cm

time, links between the heart and the soul were lost. Mechanical principles and the doctrine of the reflex could explain many physiological and psychological experiences. The impact on the physiology of the passions was two-fold. Firstly, the pineal gland, situated in the brain, became the site of interaction between mind and body. Bodily sensations were represented to the mind, where the cognitive process took place, and passions like love or anger were perceived. The same process was responsible for the physical effects associated with specific emotions, such as blushing in shame or paling in fear. Instead of being impulses of the mind/soul that were subsequently felt in the body (and around the heart, as we saw earlier in the case of the 'boiling' of the heart in

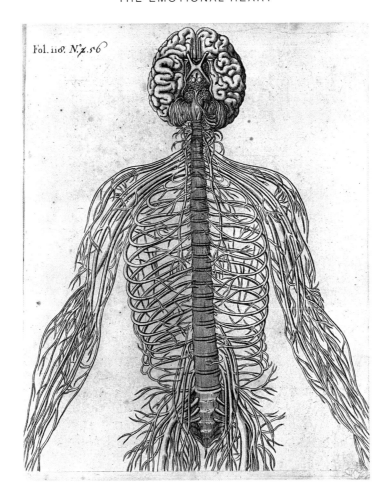

9 Diagram of the brain and nerves, in René Descartes, *De Homine*, folio 118, fig LVI, published by F Moyard and P Leffen, Leiden, 1662, engraving

rage), emotions were reinvented as bodily events, felt through the senses and transmitted to the mind. Emotions were no longer produced by the humours, or sited around the heating and cooling of the heart. Instead they were mental perceptions of bodily events that were realised through nervous transmission and cognitive processes linked to the brain.

Head over Heart: Re-Siting Emotions

It is important to note that not all interpretations of the heart in science were secular or scientific. Eighteenth-century animistic interpretations of the body continued to emphasise the role of the soul in motion and emotion. In the main, though, the period saw a drive towards the secularisation of the mind, as seen in the associationist theory of British philosophers like John Locke (1632–1704). In *An Essay Concerning Human Understanding* (1689), Locke rejected the traditional belief that individuals were born with innate principles and ideas. Instead, he described the mind as a *tabula rasa* or blank slate, to be written on by experience rather than being inscribed from birth.

At the same time as eighteenth- and nineteenth-century scientists, philosophers and intellectuals began to focus more on the role of the brain in cognition, memory and identity, the heart became subject to scientific investigation and experimentation. There were many reformulations of emotions and the mind/body relationship from the seventeenth century, including the emergence of hydro-dynamical principles in the eighteenth century, and nerve theory in the nineteenth century and beyond. None of these attributed emotional status to the heart, focusing instead on the structure of the body and its solids, and communication between the nerves and the brain. The onset of scientific medicine proper from the nineteenth century, with its emphasis on quantifying and standardising the body's systems, meant that the heart became subject to more intense medical analysis than ever before.

Under the influence of French physicians like Théophile Hyacinthe Laennec (1781–1826), discoverer and populariser of the stethoscope in 1816, and Jean-Nicolas Corvisart (1755–1821), personal physician to Napoleon Bonaparte, the heart and its diseases were described, quantified and classified. In Corvisart's *Essai sue les maladies et les lésions organiques du coeur et des gros vaisseaux* (1806), we have the first systematic classification of heart disease in clinical terms and according to new scientific principles. Cardiological techniques focused on the heart as an objectified muscle, subject, like any other, to disease, dysfunction and decay.

A parallel rise in the disciplines of psychiatry and the neuro-sciences saw the emotions being increasingly linked to the secularised brain. From Charles Bell's (1774–1842) *Idea of a New Anatomy of the Brain* (1811), to François Magendie's (1783–1855) *A Summary of Physiology* (1822), neuro-physiological investigation followed Descartes's work on the reflex in uncovering the operations of the brain and the central nervous system. In the terms of clinical science, the heart felt the consequences of emotional states – as demonstrated by palpitations in fright, or a heightened heart rate in anger. But this was simply a reflection of nervous or electrical communication between the brain and nervous system on the one hand, and the heart and cardiovascular system on the other. It did not reflect any emotional qualities or experiences in the heart itself.

At the level of popular and cultural belief, however, the emotional heart remained intact. In the eighteenth and nineteenth centuries, the time when scientific rationalism was supposedly taking hold of the hearts *and* minds of educated British people, there was a flowering of interest in the romantic heart, crystallising in such 'religions of the

ESSAI

SUR

LES MALADIES ET LES LÉSIONS

ORGANIQUES

DU CŒUR

ET DES GROS VAISSEAUX;

EXTRAIT DES LEÇONS CLINIQUES

De J. N. CORVISART, PREMIER MÉDECIN de LL. MM. II. et
RR. Offic.ʳ de la Lég. d'honneur, Professeur hon.ʳᵉ de l'Ecole
de Méd. de Paris et du Collège imp.ⁱ de France; Médecin en
chef Adj. de l'Hôp. de la Charité, Médecin cons. du I.ᵉʳ Dis-
pensaire, et Membre de la plupart des Soc. sav. de la France.

PUBLIÉ, SOUS SES YEUX,

Par C. E. HOREAU, Docteur en Médecine, Chirurgien
des Infirmerie et Maison de l'Empereur et Roi.

DÉDIÉ A L'EMPEREUR.

Hæret lateri lethalis arundo.
VIRG., Æneid.

A PARIS,

DE L'IMPRIMERIE DE MIGNERET.

1806.

10 Title page of Jean-Nicolas Corvisart, *Essai sur les maladies et les lésions organiques du coeur et des gros vaisseaux*, published by Mequignon-Marvis, Paris, 1806

heart' as Methodism and the cult of Saint Valentine. The status of the heart as symbol of emotion, knowledge and truth in the modern age is addressed by several contributors to this volume. Why was it that popular and cultural representations of emotions became more intensely fixated on the heart at the same time as medico-scientific theory rendered the heart *un*emotional?

It is arguable that these two trends are related; that any ruling theory gives rise to alternative explanations. This would explain the rise of Romanticism, for instance, when Enlightenment rationalism was at its peak. Could it be that medico-scientific theory was unable to accommodate lived experiences of *heartfelt* emotions within modern interpretations of the supremacy of the brain and the mind/body relationship? The increased popularity of alternative (now known as 'complementary') therapies, and the rise of psychotherapeutics connected with embodied memories, lend weight to this interpretation.

Moreover, it might be that in the future, scientific theory will shift to accommodate popular beliefs about the emotional heart. In 1991, Dr J Andrew Armour, one of the early pioneers in neurocardiology, introduced the concept of a functional 'heart brain' to explain the network of neurons, neurotransmitters and proteins that exist in the heart, as in the brain. The appropriation of this scientific language to explain the experiences of the emotional heart has been seen in recent media coverage of heart transplants. Families of some donors and recipients of transplanted hearts have claimed that not only the organ, but also the memories, experiences and emotions of the deceased have been transmitted to a living recipient. Armour's concept of the 'heart brain', as well as the notion of 'cellular memory', have been useful to many such individuals in providing a material explanation for this controversial phenomenon. Thus it may not be long before the 'heart of emotion' is returned to science: a heart conceived not merely as an organ or pump for the circulation of blood but, like the brain, as an intelligent organ that processes and retains our emotions.

Martin Elliott

Consultant Cardiothoracic Surgeon, Great Ormond Street Hospital for Children, in conversation with Melissa Larner

MELISSA LARNER: Did you always know you wanted to be a heart surgeon?

MARTIN ELLIOTT: No. No one in my family was in medicine, but the only things I could do at school were biology and French, and biology seemed like a better bet. So I went off to medical school, survived that, and then realised that I was interested in surgery. I taught anatomy for a year, and then did general surgery in my training, got bored with that, and then I decided that plastics was what I wanted to do – burns to hands and faces – reconstructive trauma. I began doing that as a career, but then I met a guy called George Alberti (now Sir George), who's a famous biochemist, and we started to talk. He made me realise that there were lots of research opportunities, and we sat down and sketched out a research plan, and we submitted it and got a big grant.

ML: And what was the research?

ME: Trying to understand why diabetics used to die during cardiac surgery, and how glucose moved around the body under conditions of stress. I'd failed biochemistry badly at medical school, so it was interesting to go back and do my MD in it. So I had a big grant, but no job. Mikie Holden, the paediatric cardiac surgeon in Newcastle, was a great guide and support to me, and helped create a senior registrar job for me in cardiac surgery. So I came into cardiac surgery with this big research idea, which we pursued.

But then, after some time doing adult cardiac surgery, I realised I was bored. I found myself clock-watching to see how long it took me to do things. I thought, 'Hang on a minute, I've stopped thinking about the patients here.' The bit that I found most challenging was understanding children with congenital heart disease, weighing up the plumbing issues, the physiology issues – and the idea that you can control some of it – that you can rebuild it and make it work. It took me back to the plastic-surgery principles that had interested me, and I realised it was more creative than a lot of adult cardiac surgery. So I started doing paediatrics in Newcastle and came down here to Great Ormond Street for a year's training. I stayed for two and never went back.

ML: When you say it's more creative, you mean you had to work things out for yourself that hadn't already been established?

ME: In those days, you had to make things up quite a lot more in the operating room, and that's really why I liked it. There are about three and a half thousand things that can be wrong with your heart when you're born – plumbing issues. And there are about two and a half thousand things you can do to sort them out. So the matrix of all those means that you never quite do the same thing twice. Obviously, as time has gone by, we've established more patterns of operations than we used to, more named operations, if you like. But babies keep getting born with things that you haven't seen before. And so you think, 'What do we do about this?'

ML: Presumably, these aren't new diseases, but they're things that babies wouldn't have survived some years ago?

ME: Yes. Before I started, operative mortality was close to 100 percent, and the headlines were 'Hole-In-The-Heart Baby Survives'. Now it's 'Hole-In-The-Heart Baby Dies – Shock! Horror!' There's a whole cultural shift, and I arrived in the middle of that change, where the mortality was still quite high, and we were finding out how to do things. Technology was improving, diagnostic imaging was improving, understanding of things was getting better, but the people who'd done all that were still around. All the people I looked up to when I started, are still around. Well, most of them. Some of them have died. Some are on to their trophy wives. [Laughs] So you could talk to these people and realise that you could actually be involved in something new and developing, and at the same time rapidly and spectacularly bring down the mortality.

ML: Did you find that there was any resistance to paediatric heart surgery at that time?

ME: No, oh no. People always want babies to get better. So I think 'resistance' would be the wrong word. But new things are always under suspicion. People want to know what the outcomes are going to be, and there's this sort of concept that if you don't know the outcome, then it's not worth trying, or that the risk is too high. Now, some of that's compounded by issues of resource allocations: 'There isn't enough money to do this.' Some is: 'We don't know what the outcome is going to be; you might be generating a handicap rather than solving a problem.' And some of it's just plain uncertainty. There seems to be almost envy about doing this technically challenging thing, as if it's just for your surgical ego that you're doing it, and not for the benefit of the child or the family. But what's very different in the paediatric world is that the family, and the child when they're a bit older, are directly engaged in any discussion in a much more up-front way than in adult practice – both then and now, I think.

ML: Why is that?

ME: I'd say the parents are tougher on their child's behalf than they are on their own. It's like when you go to the school with your child to have a chat with the headmaster or headmistress, you become demanding if things aren't working out at the school. You don't put up with bad things; you sort it out and you do it because it's your child – it's what you have to do. So in a children's hospital, all the parents come well-informed, prepared to argue, prepared to discuss. And that's fantastic for us. You're in much more of a partnership; it's much more egalitarian. It's much harder to be, you know: 'I'm a doctor, I'm a doctor, and I want my sausages!' It's much harder to take those sort of pin-stripe-suit approaches to medicine. It's much less patronising, in essence. I've got psoriasis, and when I go to an adult hospital now, I'm horrified by how little change there's been in the last twenty-five years in the doctor-patient relationship and the way in which ward rounds are done, for example. You don't see any of that here. It's just 'Let's sit down, have a chat, see what we can sort out.'

ML: When the children are a bit older, do you talk to them directly?

ME: Yes, of course. Although the vast majority of children we see are newborns, or within the first six months of life, so the conversation with them is limited [laughs]: 'How was your milk today?'

ML: With the older children, what strategies do you use to communicate clearly about their condition or treatment without frightening them?

ME: You just talk to them. Children are incredibly robust. It's usually the parents who don't want you to talk to the child. Children are usually so curious, and will want to know. You bring a model heart out and stick it in front of them and they say 'How does that work? Is that what mine's got?' Or, increasingly, I use video or pictures to show them things. It depends on the child; some just like a verbal description and some like to see a few diagrams, some to feel a model, others to watch a moving image. We don't have a specified mode of delivering the information, but the children will definitely be involved. But the age at which you can generate that involvement will be infinitely variable. In my experience, that can go down to as little as four, easily. We say, 'This is what you've got, this is what we do about it, this is the risk, what do you think?' and they'll say, 'I'd fix it.' It's *how* you say it. You can't be patronising, but you can tell the truth.

ML: At what age does the child start going to an adult hospital, and how do you know when they're ready?

ME: Sixteen to eighteen, depending on the maturity. There's a bit of flux either way. Obviously, you've got some people who are disabled in some way or other and aren't really ready to cope in the adult world, so we hang on to them a bit longer; others go a bit earlier. We have clinics that are devoted to judging that timing or transfer.

ML: If you were working on their livers or kidneys, do you think you'd have the same kind of relationship with your patients, or is the heart a special organ that means more to both patients and doctors?

ME: I think both of those statements are true. I hope I'd be exactly the same if I was operating on their big toe. I ought to be. The people I see working on those organs in this hospital are absolutely the same. It's something about the institution as well. Great Ormond Street is often described as a 'family' or 'home' by both the people who are working here and the patients, and that's how it feels. So you enter this partnership, wherever you are. Now, as to whether the heart has some particular *value* . . . Certainly for me it did: there was drama associated with it and I've always been the kind of person who likes to hang ten toes over the front of the surfboard or go skiing a bit too fast.

ML: You seem to share that with some of the other surgeons I've talked to. Is there a certain personality, do you think, that goes in for this profession?

ME: Yes, it's a Type-A personality. It's all a bit lunatic. A bit of danger's good for us. We cope well with the stress of emergency. We don't cope with other stresses particularly well. Emotional stress we're good at. I don't think we cope well with organisational stress, and I don't think we cope well with idiots. But I think what we are pretty good at is acting quickly on a relatively limited amount of information, in a decisive way. There's something about cardiac surgery and surgeons that means you have to do that. You have very limited time to do the work. You often don't have enough information – although less so these days than when I started – but you need to react relatively quickly and *do* something and take authority in an environment that you'd probably have been intimidated by just a few years earlier. For periods of your day, you're in control of something, and for those periods of time, you're the authority figure, and that must be attractive to certain types of people. And some of that's just to do with the decision-making pace.

I think over the last ten years, there's been a personality shift in surgeons. I suppose we'd categorise ourselves nowadays as 'post-cowboy' – perhaps 'cowhand'. The cowboy was the, you know, 'I'm a cardiac surgeon, I can do anything. Give me the knife, this child needs a new heart!' You know, the Barnards, the Donald Rosses. All of those people were from an era where nothing had ever been done before – they were making it up; they did some stuff in the lab and then they thought: 'That'll work, let's go and do it in a human.'

ML: But these people are real heroes, aren't they? They're the pioneers, perhaps, as opposed to 'cowboys'.

ME: I know, but there are two definitions of the word 'pioneer', one of which is 'maverick'. That, I think, is close. Pioneer and maverick overlap. Walt Lillehei was always called a 'maverick'. I think there's some element of that in all those personalities. We came in a little bit later. Nowadays, your results have to be perfect from day one. So you're finding a shift to a more obsessional personality: people who are great at attention to detail. We have a lot of attention to detail, but also a more adaptive style. A lot of the people I see coming through now are less adaptive, but very obsessional; getting excellent results – predictable, excellent results. So something's changed during that time. Different people have different specialities. And it's more technical and less emotional. In a sense, it's less creative, although I'm sure there are areas of creation going on, but they're more localised than they used to be when the whole thing was a drama. Now cardiac surgery is, you know, 'What's on the cardiac list? Oh, just a transplant.' The cycle of work is so heavy, with pressure from above, administration, the government; it's all filling your day, so the emotional side is being taken away from you. The job satisfaction to some extent slips away as the bureaucracy increases. A combination of obsessional personalities, greater attention to detail, a slight increase in the remoteness of the emotion, are threats to the job satisfaction of the cardiac surgeon, or they are for people like me, whose personalities probably need a bit of feeding on patients saying 'thank you', or on just seeing them smile.

But the magic of the heart, the sort of, 'Oh my god, there it goes again!', is always there. You do the operation and the bloody thing starts up and goes. Ah, it's fantastic! You never get used to that.

ML: What's it like doing a heart transplant?

ME: Well, usually it's the middle of the night so all you want to do is go and have breakfast!

ML: But what does it feel like when you take it out of the chest?

ME: Scary. But the more you do, the less scary it is.

ML: It must be incredibly thrilling as well, when the new one begins to beat.

ME: Yes, incredibly so. It's wonderful. I've made some videos that we share with people who've had transplants. You sit down with them and you say: 'This is yours, going in the bin, and here's the new one arriving.' And everyone's going 'Ooh, aah', and you feel it yourself. It's amazing. I think I always feel that my job is an immense privilege, a stroke of luck. I fell through the right number of open doors at the right time in my career, and found something I like doing and *can* actually do. And also, I think in a place like this, you're not just doing it, you're doing it in competition all the time, with your ten toes over the front of the surfboard – you're bench-marking against the best people in the world in your era. And to find yourself in that field is a shock. I feel so surprised to be here. And you think to yourself, 'This isn't quite what I intended in my life, but I'm happy with it.' It remains a challenge, it remains intellectually stimulating. There are still new things to do. And the magic of the heart isn't just about the stopping and starting, old and new. It's the fact that it's what you need to keep going. It happens quietly. It pumps away. It does its stuff all the time. It's not the 'seat of the soul' type thing – it's the magic that this bloody thing keeps going irrespective of what you do to it.

ML: Except when it doesn't!

ME: Yes. Except when it doesn't. That's where we step in.

ML: And when it does go wrong, when a patient dies – how do you cope with that?

ME: Well, fortunately, death is rare now. Mortality here is 1.8 percent. Death in the operating room – surgically related death – is almost zero.

ML: That's extraordinary. Even if you're doing a transplant?

ME: Yes, we haven't lost a transplant for six years.

ML: And is a child likely to have a full life after a transplant?

ME: No. There's an actuarial survival decline. It depends on the diagnosis, but we're about 85 percent confident that they'll be alive in ten years.

ML: That's still very good.

ME: Yes. Better than death – most of the time. I think that dealing with death now is a well-managed process. It doesn't make it easy, though.

ML: You mean there's a procedure at the hospital?

ME: Yes. It's a team thing.

ML: Francis Wells said that he felt he had that team support when actually operating, but he also talked about a certain loneliness when it was all over, because really only another heart surgeon is going to understand.

ME: I think it's special to children's hospitals, because you're supporting the family. The whole thing is about supporting the family, and each other. The first thing you do is cry. The second thing you do is put your arm round the nearest nurse, and everybody's hugging. Then it's caring about the families, putting the child into an environment where, hopefully, it'll die in the mother's arms, being there if necessary. Whatever they want, you do. And your job, very often, is to spread that process out over a period of time that's long enough for the family to feel that they've shared in the process of death. The sudden death that used to occur when I started, the deaths in the operating room or bleeding to death during the night, which still occur in adult practice, are terrible. And you just don't see them so often now, fortunately, but you never get used to going back into the room to see the parents. They just look at your face and they know — often you don't have to say anything. And you're knackered. It's usually been eight, ten, twelve hours. You're exhausted — you've given your absolute maximum energy to try and keep that child alive. Then you come out, and you're sort of half wondering what type of people they are, even though you've met them and you know them, because some people get violent and want to kill you. The most humbling thing is when they say, 'Thank you very much for doing all you've done.' And you think, 'Well did I make a cock up? Did *I* do this?', which, of course, is what surgeons think first of all. Your first thought is: 'What did I screw up?', and you're going through it all in your head: 'Would somebody else have done it better? Could I have called someone? What didn't I know? Why didn't I know it?' All of that goes through your head. And you're going through all that, and then, one of the worst parts about losing a patient — and this is where I agree with Frank about the loneliness — is that no one at home knows what you've been through. And they say, 'What sort of day did you have?', and you say, 'Oh, I lost a baby today.' They're very sympathetic, but you can't do it to them, or you've screwed up their day, big time. So you think to yourself: 'I'm not going to get into all that. I'm going to internalise it all.'

And then the next stage is that we now have deeply analytical meetings about what went wrong. We have a no-blame culture here, which is very important. All our meetings are handled with red cards and yellow cards, so if you start allocating blame, you get a yellow card. If you get a red card, your operating privileges, or whatever you happen to do, may be taken away from you. So we're very tough on it, and it's made a big difference to the way we debate things. We look at the sequence of errors that led to a death, or the lack of information about the child that led to a death, or the fact that we did everything and it was unavoidable. The team functions well, and the more we analyse these things, the better we are and the less risk there is.

ML: Do you think it's harder now to make the sort of pioneering changes that were made in the 1950s?

ME: Well, if I was trying to pioneer something, it wouldn't be in cardiac surgery, unless it was in the lab, making tissue engineering or trying to grow a new heart or a new bit of a heart.

ML: Would the NHS actively stand in your way now? Terence English said that he thought some manager would probably stop you now.

ME: No, we wouldn't let it. We'd go round it. And we *are* pioneering stuff. Most of it is to replace surgery with something else like putting heart valves in through the groin.

ML: What do you think is the future? Will it be artificial hearts, animal hearts or stem cells?

ME: Yes, yes and yes. But we can't work out which is going to dominate. We have artificial hearts that we put in now, but they're not exactly portable. They're not exactly long-term strategies, because you could have a stroke while you're on those things. There's miles and miles to go. They're really just stop-gaps to transplantation at the moment. The destination-therapy artificial heart is sort of end-of-life stuff, where they'll put it in and it'll keep them going where they would have died and any extra years are better than no years. So that won't come about very quickly. Heart replacement through tissue engineering or through animal hearts is much more subtle, and parts of hearts will be in within five years.

ML: It must be even harder to find a donor heart for a child than for an adult.

ME: Yes, we struggle all the time with it and we never have enough.

ML: How do you make these difficult decisions about who gets one?

ME: We do it as a group. This team, particularly, is very successful at working together. We argue, we record the debate, we come out with a conclusion and then stick with it. And then we review that at another meeting a few weeks later and we do the same thing again, and if we don't agree, then we revise our decision or ask for more investigations. So you have an internal peer review of your decision-making process.

ML: So no one person has 100 percent responsibility?

ME: Someone is always accountable from a legal point of view, but in reality that kind of responsibility isn't practical any more. There isn't a completely right answer to most things. There's a nearly right answer and a nearly wrong one, and you have to find some middle ground that's acceptable. First, we try and sort that out without the parents, because *we* need to have a view, and one of the most difficult things we have to deal with now is unreal expectations. People arrive thinking that their child is completely correctable because that's what we, as doctors, have been telling them for twenty-five years – we can do anything. Well, we *can* fix anything, from a technical point of view. But whether it's *right* to fix it is a whole other question. If you've got a child who's got multiple anomalies, weighs about two kilos, has pretty bad heart lesions, a single kidney, and is going to go into renal failure in five month's time, then yes, I could fix the heart. But should I? So then, how do you debate this with a family? You've got to get to know them a bit; if you don't think it's worth treating the child, you've got to be firm enough in that view to say that if it was your child, you wouldn't treat it. Then you've got to think, 'Would my wife agree with me if I said that?' Which is a slightly different question. Then you've got to ask yourself, 'Do I feel strongly enough about this to impose my will on the family?' And who has the ultimate right to decide the fate of that child? An Islamic person might take the view that the child's illness was God's will, and that I've been put on earth to have a go at saving their child, and if it doesn't survive, at least I had a go. Other families would say, 'You have a go and if you don't make it, it's your fault.' In other words, I took a risk that I shouldn't have taken. Other families will say, 'It says here in the newspaper that you can make my child well. Why haven't you?' So, managing expectations, being completely honest about what the outcome is and getting the family to accept that, when all they want is their baby to live, is a real challenge in modern medicine. And then, when a child doesn't survive, or is dying, what do you do about death? Your problems now aren't about acute death, they're about chronic death.

ML: Can you explain the difference?

ME: I'll use an extreme example. Let's say we get talked into doing a terribly complicated procedure on a child by the family. We advise them not to do it, but they, and maybe their referring physician, say 'No, give it a chance.' And we all talk about it and we're persuaded – we shift from 50/50 to 51/49 in favour of doing the operation. We do the operation and, guess what? It doesn't go very well and the child is on a ventilator, on a huge number of drugs, and is just hanging on by the skin of its teeth. And intensive care is so good now that we can keep it hanging on by the skin of its teeth for quite a lot of weeks, but then it dies. In the old days, families would come to us and say, 'My child is suffering, we're suffering, we want you to withdraw treatment.' We wouldn't argue; we'd just withdraw the treatment. Now, many parents, and particularly parents who come from a variety of extreme religious backgrounds – extreme Islamic, extreme Jewish, extreme Christian – have a view that that is completely immoral and wrong, and that life is the most important thing, irrespective of its quality, and that we must never withdraw or reduce treatment. The law protects that view, usually, and in order to challenge decisions, we'd have to go to court and we don't want to do that. So our ICU is sometimes full of people who are slowly dying. Our problem now is a resource one: it's an opportunity crisis. I can't do other stuff because those children are there, and I don't have many beds. Our team, our nurses particularly, have to nurse this child whom they know will die, and get to know this child day after day – they know it in many ways better than the parents, who actually slowly withdraw; they don't want to withdraw, but they're not there. They only come in for the case conferences, when they express very, very strong views, often with an adviser. And the poor nurses have to watch this baby die. So we have a rising level of stress in our staff – a big social problem – and we're emotionally unprepared, for the 'ten-toe' reason, for long-stay patients. We self-select to look after the patients who get better or are in the ICU for ten days; it's what we're good at. So suddenly, you're tested on the edges of your knowledge base, you're tested on the edges of your emotional base, and the cultural environment in which you're working has changed to one that doesn't have a whole lot of logic to it. And the question the nurses always ask is, 'Why have we forgotten that we're here for the baby, not for the family?' I don't know how you rationalise arguments that, on the one side would say, 'These are young parents whose baby is going to die; we should encourage them to understand that and they might be able to go away and have another baby', or on the other hand says, 'Wait for six months', where the extra trauma maybe interrupts their relationship or stops them wanting to have other children in the future. These are really difficult emotional issues that are totally different from what I thought I was getting into, which was about life or death, not life and long death.

ML: What would be the one single thing that would improve what one could do to help people with serious heart problems?

ME: Getting rid of surgery would be the most important thing. Developing non-surgical ways of doing stuff. People don't like having operations. I'd put the developments in cardiology high up there. Otherwise, I'd say that what would make it better would be enough facilities. They're constantly under threat in the UK.

At a more abstract, scientific level, what you want to be able to do is to model the procedure in advance. So I'd be looking for three-dimensional construction techniques that allow you to practice in sufficient detail, to know what you're going to see and then post-op to be able to overlap those image sets so that you'd know whether you've made a mistake.

The other thing would be all about teamwork and how teams interact: teams working well together at diagnosis, teams working well in the operating room, teams working well post-operatively, sharing information, and the best people at a particular time being allowed to do what it is they're best at. Surgeons used to move up the ladder to being the number-one surgeon and stay there, but you can't be good at everything all the time; at various stages in your career, you're better at one operation than another. And I see my job now, as generating a portfolio of people who are able to do everything, and to try and recognise individual's special skills in a particular era, which might only be for six months, but at that moment they're doing it perfectly, so you want them doing mostly those operations.

ML: So you get into a sort of a zone as a surgeon?

ME: Yes, it's just the same as anybody else. With soccer players, they play well for a season, they go off for a season and they come back again. Cricketers too. There's no real difference. It's partly a craft, partly an art, partly a sport. It's certainly a team sport. So teamwork is right up there at the top, including respect for your colleagues, and non-competition with alternative therapies. In this country, whether a cardiologist fixes a heart valve via a catheter, or a surgeon fixes it, doesn't induce much anxiety

– it doesn't threaten my existence if someone finds a better way of doing things. In the United States, it's a huge problem: the cardiologists 'steal' the work because they're financial competitors. They say, 'You don't need a surgeon; I can do it.' So the surgeon's income goes down, and the cardiologist's income goes up, and since the cardiologist is the gatekeeper who sees all the patients first, they're not going to send that patient to surgery. Surgeons blame the cardiologists – although people don't want surgery anyway, so it's a self-fulfilling prophecy – but the surgeons are still hacked off. They blame their colleagues. Immediately, the teamwork breaks down. There's a real advantage in the equity of pay in the UK and the fact that we don't have a procedure-based system of pay. So I'd put teamwork right at the top of the things I want to see happen, as well as pre-operative reconstruction and a strategy that says that surgery has had its day.

The Transplanted Heart
Surgery in the 1960s

Ayesha Nathoo

The first human-to-human heart transplant, conducted in Cape Town on 3 December 1967, was an international medical and media phenomenon. Like the moon-landing two years later, this operation has symbolised scientific and human achievement ever since. It was instantly reported as a success, a historic accomplishment, transforming Christiaan Barnard, the dashing surgeon who led the operation, and his fifty-four-year-old patient, Louis Washkansky, into celebrities overnight. The procedure is now common: around 3,000 heart transplants are conducted worldwide each year and many patients return to active lives and survive for over ten years. Yet between the first momentous operations and today's routine surgery, heart transplantation was all but abandoned for a decade. Although Washkansky lived for only eighteen days with his new heart, during the following year over 100 operations were performed worldwide by forty-seven different medical teams. But by 1970 the procedure was essentially brought to a halt.

How did 'quite a simple plumbing job', as a pioneer surgeon recently described it, become one of the most famous and controversial operations of the twentieth century? And why did human heart transplantation commence, then come to this abrupt end in the late 1960s? To explain these crucial shifts in the history of heart transplantation, one cannot think in medical terms alone. The technical advances that made transplanting the heart feasible, and subsequently severely questioned, went hand-in-hand with conceptual, institutional and cultural changes. The body had first to be seen in terms of an assortment of 'spare parts', and the heart, with its unique cultural symbolism and inextricable links with notions of personal identity, had to be conceptualised as a pump that was not only repairable but replaceable. Heart transplantation

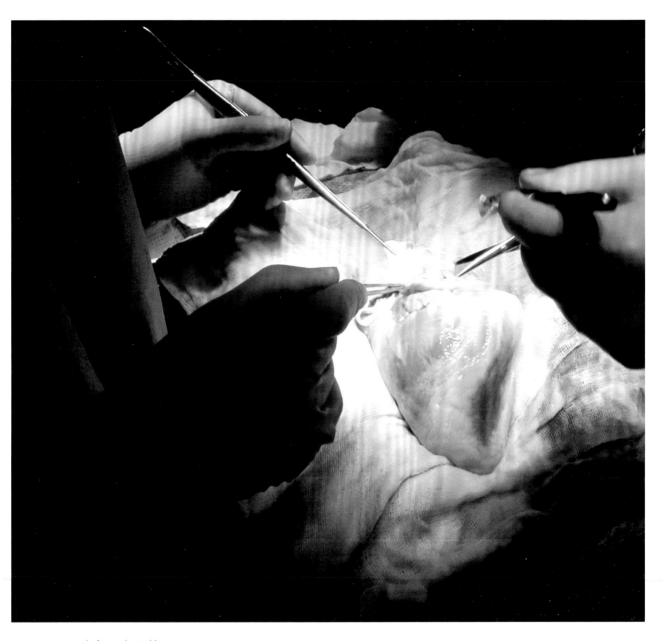

1 An explanted heart

also required a pioneering attitude possessed by daring surgeons in the post-war decades, and a belief in the potential of high-tech medicine. Moreover, since for centuries the heartbeat had demarcated life and death, heart transplantation necessitated a fundamental re-evaluation that defined the absence of brain activity in potential donors, rather than pulse, as the new signifier of death. Most of the early heart-transplant recipients died within days or weeks of their revolutionary surgery, some within hours. But these statistics alone cannot explain the moratorium: numerous other pioneering surgical procedures have had equally high initial mortality rates, yet continued unquestioned. The unprecedented media attention that firmly placed the heart-transplant controversy in the public arena seems to have been decisive.

Preparing to Transplant the 'Pump'

It was in the late nineteenth century that surgeons seriously started developing the idea that defective internal organs could be replaced. At the same time, physicians were reconceptualising the heart in terms of its functional capacity as opposed to its anatomical structure. In the early twentieth century, the French surgeon Alexis Carrel (1873–1944) and the physiologist Charles Guthrie (1880–1963) conducted hundreds of experimental organ transplants on animals, after devising a method for suturing together blood vessels. Their seminal 1905 paper, 'The Transplantation of Veins and Organs', made reference for the first time to a 'transplanted heart', and in 1907, Carrel described his technique of heart transplantation in a dog. In these early experiments, the transplanted hearts were placed into the recipients' necks and abdomens, mainly in order to improve physiological understanding, but Carrel also tentatively speculated about eventual therapeutic benefits. However, he foresaw significant problems in acquiring and preserving suitable donor organs, and believed that the future of trans-plantation therapy in humans depended on using animal organs. Anti-vivisection groups vehemently opposed his animal experiments, but the greatest barrier was that the recipient bodies invariably rejected the transplants.

In the 1940s and 1950s, this rejection process became understood in terms of the body fighting the grafts as it would a disease, in other words, identifying the new organ as 'not-self' and therefore provoking an immunological response. The work of the British-Lebanese professor of zoology, Peter Medawar, on the rejection of skin grafts for burns victims was foundational to developing the field of transplantation immunology. The first successful kidney transplant was performed in 1954 in Boston on identical twins. The brothers' identical genetic make-up meant that the trans-planted kidney was not rejected, and the patient, Richard Herrick, survived for eight years. The first kidney transplants from unrelated living donors and then cadavers were decidedly less successful, as were the early attempts at human lung and liver transplantation in 1963. But despite high failure rates, renal transplants continued experimentally in the early 1960s. Longer life expectancies were achieved, and the procedure was safeguarded by dialysis machines that could be used as a long-term back-up to unsuccessful transplantation. No equivalent device existed for the heart, and human heart transplantation remained, even in the early 1960s, an undertaking that most doctors, like the general public, thought could not, and perhaps should not, be performed.

Experiences in the First World War had encouraged greater surgical experimentation and therapeutic confidence, and led surgeons to attempt increasingly audacious operations. Hysterectomies, lobotomies and radical mastectomies were subsequently performed in their thousands. Cardiac surgery, however, was mainly a product of the Second World War. Even in the early twentieth century, surgeons believed that they could not touch the heart, due to its role of maintaining life. Experimental surgery on Second World War injuries, however, facilitated by developments in antibiotics, new imaging and measurement technologies, as well as advances in blood transfusion, proved for the first time that the heart could be interfered with and repaired.

In the 1950s and 1960s, the nature of surgical intervention shifted from removal of tissue parts to restoration and replacement. Cardiac pacing and defribrillation were indicative of this new phase, as well as valve replacement. Growing in professional prestige, surgeons were at the forefront of scientific advance, often compared to the new astronomical explorers, and credited with using recent technologies for humanitarian, beneficial means, rather than for weapons and destruction. As a *Time* magazine feature declared in 1963:

> Under the bright lights that illuminate the surgical incision with brutal clarity, the achievement of the surgeon and his assistants becomes one of the greater glories of science. Man may strain ever farther into space, ever deeper into the heart of the atom, but there in the operating room all the results of the most improbable reaches of research, all the immense accumulation of medical knowledge are drawn upon in a determined drive toward the most awesome goal of all: the preservation of one human life.

After the Second World War, as heart disease took over from tuberculosis as the greatest threat to Western lives, thoracic surgeons turned their attention to cardiac surgery, and tuberculosis centres were increasingly adapted into cardiac units. Cardiac surgery, dealing as it did with matters of life and death, epitomised the image and attitude of the 'heroic' surgeon, and the heart's unique symbolism added to its appeal. As one doctor professed in the late 1960s: 'Cardiac surgery . . . is the most intensely dramatic [surgical speciality] due in part to the wide variety of intricate operations and the complicated apparatus needed, and in part to the emotional symbolism of the heart, rooted deep inside us all.'

The heart-lung machine, devised in the early 1950s, which took over the circulation and oxygenation of the blood, was a crucial innovation for making open-heart surgery technically feasible. The machine's alternative name, the 'pump-oxygenator', indicated the mechanistic conceptualisation of the heart, seen in terms of its function as a pump, which in principle could not only be repaired, but also replaced. Artificial, animal and donor human hearts all became potential candidates as substitute pumps.

After the successful application of the heart-lung machine, which seemed to indicate that the heart's function could indeed be substituted mechanically, creating an artificial heart appeared all the more viable. Work had commenced in the 1940s and 1950s, but the project came to fruition in 1963 in the United States with the launch of an artificial heart programme sponsored by the National Heart Institute. Following other highly co-ordinated post-war projects, the government, public and researchers were confident that with the application of sufficient funding and institutional support, great minds would be able to solve what seemed to be just a complex engineering problem: to create an artificial pump that could simulate the function of the human heart. The initial target was to develop a fully implantable, total artificial heart by Valentine's Day 1970. This choice of deadline, a day most closely identified with the heart's cultural and symbolic significance, manifested the complex concurrent meanings and associations of the organ.

Rather than striving towards artificial replacement, which proved to be significantly more problematic than anticipated, other researchers saw greater scope for therapeutic success in transplantation. Yet, even in 1951, leading cardiac researchers considered that human-to-human heart transplantation was 'at present a fantastic dream'. By 1964, surgeons, mainly in the United States, Russia and Britain, who had persevered with animal experiments in the 1950s, were claiming that this fantastic dream was now 'just around the corner'. In January 1964, the Mississippi surgeon, James Hardy, who had performed the world's first lung transplant the previous year, was ready to conduct the first human-to-human heart transplant. In the event, the critically ill recipient went into terminal shock before the potential donor's heart had ceased to beat. Not prepared to switch off the donor's ventilator, Hardy transplanted a chimpanzee's heart into his dying sixty-eight-year-old patient, but the small animal heart was unable to cope with the circulatory load of a man, and he died after an hour. Although the procedure was unsuccessful, by this time, the surgical and technical know-how to transplant a human heart was clearly in place. Nevertheless, later that year, Norman Shumway and Richard Lower, who had led the way in cardiac-transplant research in the United States, suggested to their colleagues that 'perhaps the cardiac surgeon should pause while society becomes accustomed to the resurrection of the mythological chimera'.

Their reservations about pushing forwards into the clinical arena were primarily social. Heart transplantation raised a number of troubling, ethical, legal and financial issues, and the pioneer surgeons did not even have the full support of their medical colleagues. Many believed that the limited financial resources would be better used for already successful cardiac treatments. Moreover, there were specific concerns about how the volatile public of the late 1960s would react to the reality of changing human hearts. It was a time of growing activism, awareness and calls for accountability, manifested, for example, in the anti-war, feminist, student and environmentalist movements. Television was the new mass medium of the period, and specialist and investigative journalism were on the rise. In this climate, medicine, too, began to be viewed as an activity, like any other, that should be open to public discussion. The thalidomide controversy in 1961, when thousands of babies were born without properly formed limbs due to an anti-nausea drug administered to their mothers, had significantly shaken confidence in the medical industry, and patients had already begun to form pressure groups and associations, pushing for 'patient rights'.

In Britain the Labour government, led by Harold Wilson, had been elected to power in 1964, advocating progress through the 'white heat of technology'. Nonetheless, in the Cold War era, an increasingly affluent, educated and consumerist society received new technologies with ambivalence. The anthropologist and Provost of King's College, Cambridge, Edmund Leach, confronted these issues days before the first human heart transplant, in his 1967 BBC Reith lectures, 'A Runaway World?'. He addressed the fear and uncertainty that accompanied a seemingly out-of-control, fast-paced, highly technologised society, where 'Men have become like gods':

> The marvels of modern technology fill us with amazement but also with dread. It was alright when the surgeons just fitted us up with artificial arms and legs, but now that there are people going round with plastic guts, battery-controlled hearts [pacemakers], dead man's eyes and twin brother kidneys, there begins to be a serious problem of self-identification . . . Am I just a machine and nothing more?

New Hearts for Old

The first human heart transplant, on 3 December 1967, made international front-page news, reported on a scale that was unprecedented for a medical event. At first, it was widely celebrated, immediately defined by the press, both popular and medical, as a success. Even if Washkansky did not live for long, the transplanted heart had started beating again in his body. He had received the heart of twenty-six-year-old Denise Darvall, who was fatally injured in a car accident that also killed her mother as the two women crossed the road. Her mother died instantly and Darvall was taken unconscious and critically injured to the nearby Groote Schuur Hospital, where Washkansky lay dying of end-stage cardiac disease. Her life could not be saved, but her misfortune amounted to changing the course of Washkansky's life, and that of medical history.

The tale of the first man with a transplanted heart made an extraordinary human-interest story, and each of the eighteen days during which Washkansky survived

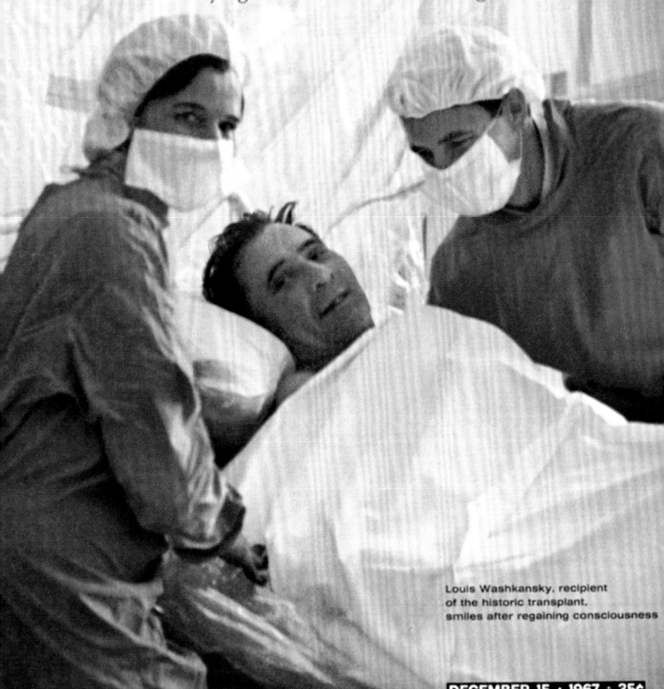

LIFE

GIFT OF A HUMAN HEART

A dying man lives with a dead girl's heart

Louis Washkansky, recipient
of the historic transplant,
smiles after regaining consciousness

DECEMBER 15 · 1967 · 35¢

3 Christiaan Barnard on cover of *Time* magazine, 15 December 1967

the operation were reported in minute detail. Newspapers headlined with the first words he spoke to his wife, what he ate for breakfast, and how he joked with the nurses. Competing international broadcasting teams flooded into the hospital, and interviews and images of the 'heart-swap man' were transmitted across the world. He became 'the most publicized hospital patient in the world'. On 15 December he appeared on the front cover of *Life* magazine, the same day as his surgeon, Christiaan Barnard, made the cover of *Time* magazine.

The South African authorities, politically isolated during the apartheid era, gave Barnard their full support, seizing the opportunity to improve their international image. Due to Nationalist Party ideological objections, South African television broadcasting did not commence until 1976, even though it had both the financial and technological capacity to introduce the medium earlier. Nonetheless, following the transplant, Barnard accepted abundant invitations from international broadcasting organisations and toured the world, meeting film stars, the President of the United States, and even the Pope. His looks, panache and willingness to deal with the press all contributed to the overwhelming media attention, but the deeply entrenched cultural symbolism of the heart was still influential. The questions posed by journalists to the patient were telling: 'Mr Washkansky, as a man, how does it feel to have a female heart?'; and 'As a Jew, what is your feeling about having the heart of a non-Jew?', asked the BBC. Meanwhile Mrs Washkansky, fearing that her husband would not love her any more, admitted: 'I was petrified at what I'd find. Like everyone else, I thought the heart controls all your emotions and your personality.'

2 Louis Washkansky on cover of *Life* magazine, 15 December 1967

Although competing with several other big international news stories, such as the drowning of the Australian Prime Minister, Harold Holt, and the enforced flight from his country of King Constantine of Greece, the heart transplant persistently made headlines. When Washkansky died on 21 December, the *Daily Mail* summed up:

> The past eighteen days have seen an unusual succession of big news stories. Among them the big freeze, the go-slows, foot and mouth, a toppled throne, a drowned Prime Minister and an acute government crisis. But one story above all has appealed to the deepest emotions of men and women everywhere. It is that of Louis Washkansky, the man with the transplanted heart.

Barnard's operation was closely followed by three heart transplants in the United States, undertaken by competing surgeons who had been on the brink of being the first, but had been beaten to it by Barnard. Two of their patients died within hours, the other within days. Barnard had meanwhile performed a second transplant, but following the deaths of Washkansky and the American patients, the medical world was divided as to whether human heart transplantation was justified or premature.

Following the first transplants, one commentator wrote in Africa's leading intellectual magazine, *Transition*:

> For centuries it has been assumed that to give your heart to someone is to pledge your love. And yet here we are in the transition between 1967 and 1968 beholding a literal transplantation of hearts with a supreme impersonality. The donor and the beneficiary are total strangers, and might never have cast their eyes on each other. The ultimate symbol of human affection is reduced to a clinical convenience.

When heart transplantation started, although donors and recipients were strangers, they were not anonymous. Patients' names and photographs were printed in the press, and their families often met. Far from the dead body being objectified and detached of meaning, families of donors often agreed to donation with the hope of transcendence – that their loved one's heart would literally live on in the new body. As Denise Darvall's father lamented after Washkansky died: 'There was at least part of my daughter still alive in Mr Washkansky. But now she is completely dead.'

Alongside the rhetoric of the 'pump', transplant surgeons used the popular notion of the heart as the 'ultimate symbol of human affection' to promote donating this 'gift of life'. This, however, also gave rise to contested ownership of the organ. For the period when Washkansky's body was accepting the new organ, was it already his? And when he died, whose heart was it then – Darvall's, Washkansky's, or simply an object of medical science? Cape Town's Chief Rabbi criticised the removal of Washkansky's second heart before he was buried, arguing that Judaism required all parts of the body to be buried intact. Surgeons in the United States and Britain had requested samples of the heart for medical scrutiny, but the organ was also needed by South African police for investigations into Darvall's roadside accident and ensuing death. Therefore, the heart was not merely an objectified 'pump': it was an important part of two different people with separate histories.

From the start, transplant surgeons had been all-too aware of the identity implications of heart transplantation. In apartheid South Africa, Barnard had been careful to find both a white donor and recipient for his first heart transplant. His second operation, in January 1968, transplanted the heart of a 'coloured' man, Clive Haupt, into a white patient, Philip Blaiberg. Exceptionally, Blaiberg survived for more than a year and a half, but, given the racial implications, news of his operation caused an inevitable commotion. In Britain, the editorial of the left-leaning *Guardian* newspaper sarcastically pointed out:

> If a white man can use a coloured man's heart after death, can he sit on the same park bench with him when both are living? Probably not. The surgeon should know that he faces grave charges. There is no provision under the Group Areas Act for black hearts to beat in white neighbourhoods. Mitigating circumstances may be pleaded for him, but there can be no doubt that Mr Haupt is committing a posthumous offence.

Much of the disagreement among the divided professional medical community on whether heart transplantation should have started and should continue, was itself played out in the media, transcending traditional boundaries between popular and specialist forums. In an article in the British magazine *New Scientist* in January 1968, a professor at Charing Cross Hospital claimed that heart transplants were being carried out because of the 'twin allures of publicity and one-upmanship'. Two days later, three eminent cardiac surgeons from the Hammersmith Hospital in London wrote a letter to the specialist *British Medical Journal* expressing grave concerns about the myriad scientific, legal, ethical and administrative problems attending heart transplantation. These articles were widely reported in national newspapers, and were closely followed by a special edition of the BBC's popular science programme, 'Tomorrow's World', entitled 'Barnard Faces His Critics'. It featured Barnard in a televised debate with a 100-strong studio audience mainly comprised of male medics. It was an unprecedented occurrence for a traditionally reticent British medical establishment. In the months that followed, more and more surgeons around the world attempted heart transplantation. Each operation attracted fresh media attention, and justifications, accusations and news of fresh heart-transplant cases continued unabated.

On 3 May 1968, surgeons at London's National Heart Hospital, led by Donald Ross, conducted Britain's first heart transplant, the tenth in the world. The donor, Patrick

Ryan, had fallen twenty feet onto his head at a building site in south London, and his heart was transplanted into the chest of a forty-five-year-old patient, Frederick West, who died in hospital forty-six days later. The publicity attending the operation itself became a major part of the news story. The National Heart Hospital for the first time held a post-operative press conference and even hired a commercial PR firm to help manage the extraordinary media attention. The naming and photographing of donors and recipients threatened notions of patient confidentiality, and doctors appearing in television studios and named and pictured on newspaper front pages broke professional codes of conduct.

The same team performed a second cardiac transplant soon afterwards, but this time the patient survived for only two days. The first two operations at the National Heart Hospital and a third and final heart transplant led by Ross, but conducted at Guy's Hospital the following May, were all embroiled in controversy. As heart-transplant patients around the world continued to die in quick succession, the initial optimistic reporting that had once depicted the operation as a provider of hope for millions of heart-disease sufferers changed into journalistic scrutiny and public alarm.

Hearts and Brains

As well as the visibly high mortality rates of heart recipients, the 'death' status of donor patients now became one of the most contentious issues. Was the donor actually 'dead' and if so, who decided, and how? British surgeons were not alone in this dilemma. During the first wave of human heart transplantation, the definition and redefinition of death became a highly public ordeal. The beating heart had for centuries been the signifier of life and death, and 'brain death' as we now know it, did not exist. The definition and timing of death had long been controversial issues, but the development of hospital-based resuscitation technologies and organ transplantation brought the matter to the fore in the 1960s. Intensive-care equipment had created a situation whereby patients' hearts and lungs could be maintained even though their brains had lost any meaningful function. Since the introduction of these life-supporting technologies, physicians faced new ethical challenges as to when to stop treating patients who had lost all reflexes and brain activity, and were trapped in a state of 'irreversible coma'. The transplantation enterprise gave this issue fresh significance, first with kidney transplantation and then with greater urgency as heart transplantation began worldwide.

The high-profile heart transplants hurried along a redefinition of death, from the cessation of the heartbeat to loss of meaningful brain activity. The most authoritative early statement was by the Ad Hoc Committee of the Harvard Medical School to Examine the Definition of Brain Death, published in August 1968. Their report characterised 'brain death' or 'irreversible coma' as unreceptivity and unresponsivity, no movements or breathing, no reflexes and flat electroencephalogram. The same month, the World Medical Assembly issued an interim statement about brain death in their 'Declaration of Sydney' that was almost identical to that of the Harvard Committee. Brain-death laws, however, were not enacted until several years later, and different definitions were used in different countries. In Britain, for example, death was redefined in 1976 as 'the complete and irreversible loss of function of the brain stem'. India did not follow until almost twenty years later, defining and legally accepting brain-stem death in the 1994 Transplantation of Human Organs Act. Despite the differences across countries, the heart transplants prompted a shift from the heart to the brain as the organ responsible for defining life and death in the last half century.

The moral, legal and religious issues associated with redefining death profoundly affected the 1960s heart-transplant enterprise. The existing definition of death in the 1960s, as the cessation of the heartbeat, raised the question as to whether taking out a beating heart for transplantation purposes would actually amount to murder. This was not just an academic question. In Japan in 1968, the transplant surgeon Wada Jiro was charged with intentional homicide and professional negligence – causing the deaths of both Japan's first heart-transplant donor and the recipient. He was accused of unnecessarily transplanting the recipient's heart and hence killing him, and not providing evidence that the donor was dead when his heart was removed. Charges against him were eventually dropped in 1972, but Wada was found to have lied to the media, and also to have tampered with the valves in the recipient's original heart to exaggerate their defectiveness. The medical and media scandal had long-lasting effects, and Japan's second heart transplant was not carried out until 1999. In May 1968 in the United States, the cardiac-transplant pioneer Richard Lower was also accused of killing a patient, in America's first interracial heart transplant. Lower removed the heart of a black man who had been severely brain-damaged following a fall onto concrete. The patient was declared 'unclaimed dead' and within hours his heart was transplanted into a white businessman, who lived for seven days with the new heart. The donor's family, who had not given permission for the removal, sued the surgeon in a case that was not resolved until 1972, when Lower was eventually cleared, and 'brain death' was legally accepted.

In Britain, accompanied by considerable media interest, the surgeons who had carried out the country's first heart transplant were also taken to court, quizzed over the moment and nature of the donor's death. At the press conference following the operation, journalists were told by Donald Longmore, one of the surgeons on the transplant team, that the donor, Patrick Ryan, had been resuscitated several times before 'it was finally decided to accept the fact that the patient was dead'. The *Daily Express* then headlined: 'Six times Patrick Ryan's heart was restarted . . . but when was he dead?' The science correspondent, James Wilkinson, queried: 'When did Patrick Ryan die? By what

DAILY EXPRESS MONDAY MAY 6 19

● Six times Patrick Ryan's heart was restarted…but when was he dead?

A NATION'S QUESTIONS
The Express presents the facts
by JAMES WILKINSON
Michael Steemson, Colin Smith,
Antony Cheesewright

THE KEY MAN in Britain's first heart transplant operation was 24-year-old Patrick Ryan. For it was his heart.

There is much curiosity about Mr. Ryan, his life and death—specially the time and circumstance of his death.

When did Patrick Ryan die?

By what yardstick was he judged to be dead?

And by whom?

Indeed, in Britain today, what is the moment of death? Two committees—one set up by the Government and another by the British Medical Association—are even now trying to arrive at definite guidance which could become law.

But until they report, and Parliament acts, death is a matter of one doctor's opinion.

It is for that reason that Patrick Ryan's death is a matter of urgent public concern.

Patrick Ryan—six 'deaths' and now a tense medical debate

4 Cutting from *Daily Express*, 6 May 1968

yardstick was he judged to be dead? And by whom?' These were framed as matters of urgent public concern, under the heading: 'A nation's questions'. At the inquest that followed, the jury gave a verdict of accidental death; Longmore remembers it as a 'very nasty' incident, 'stirred up by the press', that could have put the surgeons in 'very serious difficulties' had he not taken Ryan's actual skull and photographs of him as evidence.

The second and third British heart transplants in July 1968 and May 1969 were also enmeshed in controversy over the death-status of the donor. Journalists championed the cause, keeping the issue at the forefront of transplant publicity. The renowned science correspondent, Chapman Pincher, wrote an opinion piece in the *Daily Express*, after attending the inquest into the death of Britain's third heart donor, headlined 'Patient still alive? Doctors must stop the doubts.' 'If the medical profession believes that such doubts do not exist, it is deluding itself', he wrote. 'They exist because the public is naturally suspicious of the secrecy under which transplant operations are being performed … They exist because the individuals who comprise the public realise that they or their relatives may be the next to turn up at some hospital as highly desirable donors.'

The ever-increasing number of heart-transplant operations did raise public fears that hospitalised accident victims could have their hearts prematurely snatched away in the interests of potential cardiac transplant patients and their eager surgeons. Some doctors also had reservations. The head of intensive care at St Thomas's Hospital in London caused a furore when he told the press at the World Congress of Anaesthesiologists in September 1968 that, 'As soon as one has a patient with useful organs one has a gang of vultures trying to snatch out these organs, ranging from the cornea to the heart.' There was clearly scope for exploitation, as one concerned *Sun* reader wrote to the paper: 'A Prime Minister could need a new heart and the "hopeless" patient available might be a tramp.' In South Africa, these were precisely the sentiments felt by much of the black population, afraid that their bodies may be used and abused by and for the advantaged white people. This was at the root of much of the disquiet surrounding Barnard's second operation, as well as the 'Tucker v Lower' lawsuit in the United States. Fears of commodified 'spare-parts' have by no means disappeared today. The illegal trade in organs is widespread, with a steady stream of organs trafficked from the poor to the rich.

The controversies surrounding the late 1960s heart transplants, and the publicity that fuelled and attended them, resulted in donation rates plummeting. This not only affected supplies of hearts, but the entire transplant enterprise, most notably the more successful and established kidney-transplant programme. More seriously, the public's faith in high-tech medicine, which heart transplantation represented, and its trust in the entire medical profession were at stake. The result was a moratorium for the following decade, formalised in some places and left informal in others. There were a number of contributing factors, including the low patient survival rates and the uncertainties over 'brain death', but the most significant distinguishing factor between heart transplants and other cutting-edge, high-risk, medical innovations was that the heart-transplant drama was played out on the public stage.

It was not until the end of the 1970s that heart-transplant surgery re-emerged as an acceptable treatment for gravely ill cardiac patients. In the intervening period, only Norman Shumway from Stanford Medical Centre persisted with the procedure. Together with Richard Lower, he had carried out much of the preliminary research into cardiac transplantation, and he quietly continued with the operations during the 1970s, and achieved increasingly successful results out of the limelight. In Britain, against the advice of the Department of Health, the heart surgeon Magdi Yacoub performed a heart transplant in September 1973 at Harefield Hospital in London, but his patient survived for only four hours. No more cardiac transplantations took place in Britain until 1979, when a comprehensive programme was initiated at Papworth Hospital in Cambridge, and a year later at Harefield Hospital. The development of the immunosuppressant cyclosporine, hailed as a 'miracle drug', is usually credited with restarting the heart-transplant enterprise. However, although it did significantly improve the life-expectancy of transplant patients, it was not introduced to British recipients until the early 1980s, after the programme had recommenced. The encouraging results from Stanford, new methods of preserving donor hearts, and crucially, a change in media reportage and public and professional attitudes towards brain death, re-established heart transplantation as a desirable and achievable therapy in the new socio-political landscape of the late 1970s.

Meera Rajan

NEWCASTLE HEALTH AUTHORITY
FREEMAN GROUP OF HOSPITALS

FREEMAN HOSPITAL

FREEMAN ROAD HIGH HEATON NEWCASTLE UPON TYNE NE7 7DN

TELEPHONE
TYNESIDE 284 3111
STD 091
EXT.

REGIONAL CARDIOTHORACIC CENTRE
WARD 23

ASH/NBM/F0089213L

5th February, 1987

Mr. M. Yacoub,
Consultant Cardiac Surgeon,
Harefield Hospital,
Uxbridge,
Middlesex.

Dear Magdi,

Meera Thyagarajan, d.o.b. 05.07.81

Diagnosis: Univentricular heart
             Absent left AV valve
             Mal position of the great arteries
             Total anomalous pulmonary venous drainage to the azygos vein
             Ostium primum ASD
             Dou ble outlet ventricle
             Pulmonary valve and infundibular stenosis
             Left Gortex systemic pulmonary shunt 28.5.86 - Mr. Hilton

I would be grateful if you could look at the angiograms and haemodynamics on
this child. She is the daughter of a local general practitioner and presented
to us with cyanosis more than 1½ years ago. Our only possible option at that
stage was to shunt her and this was carried out by Colin Hilton on 28th March,
1986. She initially improved but subsequently has gone into a lot of heart
failure. This is associated with ventricular dysfunction and AV valve
regurgitation, both of which have been confirmed ultrasonically. I showed her
case to Colin Hilton and Chris McGregor and asked whether she was suitable for
transplantation. I had initially thought that this would need to be heart and
lungs but there is a suggestion thats it might be possible to do heart only
which would be infinitely preferable. On medication she is managing to lead a
relatively normal life at the moment but I think you would agree that the
combination of lesions and the fact that she has been in heart failure suggest a
rather limited outlook. Father would be very happy for us to go ahead at the
appropriate time with transplantation and I think he appreciates the problems.
The family will be resident for the forseeable future in the United Kingdom.
If there is any further information that you need, I would be very happy to
supply it.

Kind regards,

Yours sincerely,

Stewart Hunter
Consultant Paediatric Cardiologist

Meera Rajan

Student, in conversation with Melissa Larner

MELISSA LARNER: What was the illness that led to your heart transplant?

MEERA RAJAN: It was congenital. When I was ten days old, they realised there was something wrong with my heart. Progressively, as I got older, it got worse and worse. From what I can remember, I had only one valve and one ventricle. Oxygenated blood wasn't able to travel around my body, which made my lips and nails quite blue. So when I was five, I had an operation on my valves. They thought it would buy me some more time, which it did, but I just got worse and worse and then when I was six or seven they said, 'She needs a new heart – there's no other way around it.'

ML: What was it like, being so ill as a child?

MR: I knew I was different, because I couldn't do things that other kids could do. I couldn't go outside to play and I got tired quite easily. But I didn't know any better because I was ill when I was born. It wasn't like, 'Oh, I can't do all these things now', because I'd never done them before. So in a way, if I had to be ill, I think I'd prefer to be ill when I was little, just because you don't know any better.

ML: When they said 'You need a new heart', how did it feel?

MR: When they took me in the room and they explained all the details, I wasn't really listening. And then when they said, 'Do you want this heart?', I remember saying, 'No, it sounds a bit complicated. No, I don't think I really want to do this.' I thought hospitals were depressing places and didn't want to be there for longer than I had to. But my parents were sitting there going, 'She means yes really.' So I'm glad my parents chose the best option available.

Letter from Dr Stewart Hunter to Mr Magdi Yacoub, 5 February 1987

ML: Without it, would you have died?

MR: Yes. I had a very poor quality of life shortly before my transplant, when I'd become very, very ill. My mum was carrying me around. She was my full-time carer; she'd even carry me to the toilet. I couldn't digest food very well – I couldn't do much really. But as I said, it didn't seem that bad at the time.

ML: Did you have any idea how difficult it was going to be to get a donor heart?

MR: No, I had no perception of that. But because I got ill quite dramatically, from being OK to being very, very sick, I got shoved up the waiting list pretty much straight away.

ML: And do you know whose heart you were given?

MR: I don't have too much information. I was meant to be part of a domino. That was the plan: I'd receive a healthy heart from someone who had defective lungs, and they'd get a heart and lung transplant. But I couldn't wait that long, so I got a donor heart from a fatality. They don't give you much information because of privacy – but I know it was from a ten-year-old boy who lived in Oxfordshire. He died in a car accident from massive head trauma. And that's as much as I know.

ML: So you didn't meet the family of the donor?

MR: No. Sometimes I wonder what it would be like if I wrote them a letter, or if I met them. But I guess it's . . . I don't know, I think if they wanted to know about me, they might have tried to contact me. It would be quite difficult for me to contact them. I think it's their choice.

ML: It's a fascinating concept that you've got someone else's organ inside you. As a child did you find that strange?

MR: No, I only started to think about it when I got a bit older. Now I think about it. It's a strange concept. There are all these stories about taking on some of the personality of the donor. I guess it'd be difficult to know as I child, because your personality is still developing. Obviously the trauma of having a transplant would change your personality as well. After my transplant I had a lot more energy. I did go a bit bonkers, but everyone said that was a side-effect of the medication.

ML: What sort of bonkers?

MR: I was just doing things I didn't do before, when I was ill, like having temper tantrums. Maybe that was because I didn't have the energy to have temper tantrums before, or maybe, because I'd spent so long in hospital, it just stressed me out in a way that I didn't really realise.

ML: Do you remember your surgeon or your consultant?

MR: Not my surgeon particularly, because they don't have that much contact with you. I remember my consultant, Dr Hunter from Newcastle Freeman Hospital. I really liked him. He was a nice guy. He took very good care of me. Everybody in Newcastle did. They were a very good team up there.

ML: Is that where you lived as a child?

MR: Yes, for the most part. We lived up North. But I remember going to Harefield to see Magdi Yacoub. I saw him once because there was an issue over whether I needed a heart and lung or just a heart. They wanted to know if it was going to affect my lungs. Now, Harefield is celebrating twenty-five years of transplantation, and all the transplant children, including me, had to walk up on stage and meet Professor Yacoub. It's quite strange: he actually remembered me. I'm like, 'I don't remember you.' I don't think I looked people in the face when I was little, because I was really shy and it was difficult having people poking and prodding at me. But he seemed to remember me, which was nice. He seemed to remember all his patients. He's got a very good memory.

ML: How long did you have to spend in hospital once you'd had the transplant?

MR: I can't remember exactly. It felt like the whole summer, or maybe longer, as I had a lot of physiotherapy. It was strange after my transplant. Because I hadn't used my legs for such a long time, and because they work on your arteries in the tops of your thighs, I was limping. And they left a syringe in your neck for the duration of your stay, which meant that my head was tilting to the right, and my right shoulder was raised. So I had to learn to hold my neck straight up, and to walk properly, and go downstairs. But even today my right shoulder is still slightly raised. I had my transplant back in '89, when they didn't know much about it, so I think they kept me in longer than they would now.

ML: It's a huge amount to go through at eight. You must have been very brave.

MR: I don't know if I was. I don't feel like I've done much! Everything was done for me, and I was just taking it in. A lot of decisions were made for me, by the people who did the operations and my parents. Those are the people who've gone through a lot. I don't remember feeling worried, because I just didn't know what everyone was talking about: 'What are you worried about? I'm just a bit blue today, it's fine.' [Laughs] But I think things would have been different if I was older.

ML: Yes, if you were a bit more conscious of death.

MR: Yes, exactly, you don't have a concept of death when you're that young. You think death is sleeping for a very long time. [Laughs] If I was older, I probably would have thought about death and I also probably would have whinged about all the things I used to be able to do.

ML: Obviously your parents were very protective, and shielded you from everything they could. But did you ever feel they were wrapping you up in cotton wool?

MR: They did. They probably still do. But that's because I'm the youngest, as well as being ill. They just worry. Don't all parents?

ML: Did you find yourself not always telling them if you had a symptom of some sort?

MR: No, I think they were pretty much aware of things, and I'd just say if was feeling ill. My father's a doctor. That almost made things worse for him because he knows everything that's going on.

ML: That must have been comforting for you though.

MR: Definitely. He was always watching and aware. Back then, transplantation was in its early stages, and a lot of the medications were quite new, and he'd notice if I was reacting badly to something. I remember once I had a bad reaction to one of the medications: I was kind of swelling up. He was like, 'I don't think that should be happening to you!' [Laughs]

ML: Having a doctor on hand must be good even now.

MR: Definitely, especially now. As you get older, they put you on more and more medica-tion, just to stop other symptoms and side effects, and you've got to be careful, because they could interact with one another. It's always useful to run to him and say, 'Should I really be taking this with everything else I'm taking?' Sometimes I think he's a bit shocked by the recommendations that some GPs give me.

ML: What kinds of medications are you on?

MR: They're immunosuppressants, to stop me rejecting the heart.

ML: So you'll be on those for the rest of your life?

MR: Yes, and unfortunately some of them have side-effects. One of the immunosuppres-sants affects your calcium absorption. I had a bone-density scan at the end of 2005, which showed that my bones were slightly thinning. It's not osteoporosis, but it means it could start. So, they put me on another type of medication that meant I could absorb calcium better. But that triggered my gout.

ML: You got *gout*?

MR: Yes. [Laughs] It's difficult, because they put you on one thing to stop the side effects from another thing and that causes it's own side-effects. When I got the gout this year, and I wasn't able to walk that much, I was annoyed and frustrated that I couldn't do things like go out with my friends or go to work. I'm one of those people who just does things constantly. I can't sit down for very long . . .

ML: Oh dear!

MR: Yeah. [Laughs] So when I was ill, I was like, 'What am I meant to do? *What am I meant to do*?' I had a lot of work on at uni, and it was quite frustrating. When I was little, it wouldn't get me down so much. When you're little, you don't really feel guilty about not doing things.

ML: How many tablets are you on a day?

MR: I guess about five or six in the evening and four in the morning. I'm used to it now. It's just my routine cocktail for the morning and evening. [Laughs] I'm just a bit high maintenance! But other than that I'm quite normal.

ML: At school were you an object of curiosity?

MR: I guess I was. And I think it's quite strange when people find it interesting. I mean, before I came to do this interview, I was talking to this girl, and I said, 'I've got to do this interview. I've got nothing interesting to say. What am I going to say?' I guess they found it interesting at school, and I couldn't quite understand why.

ML: But it is interesting. When you say to your friends at university, 'I had a heart transplant at the age of eight', they're not going to shrug and say 'Oh really?' They must say, 'Oh my god!'

MR: They do. They say 'Wow!', or 'Oh my god!'

ML: A heart transplant is a big thing to go through. And that's why people are interested. We don't experience people tinkering around inside our bodies very much, and we're scared of it. Especially if it's the heart, which is such a crucial organ. I had a tooth transplant when I was little, and people thought that was weird!

MR: [Laughs] You know, I guess it is strange, but this is just the little world I live in. I've just grown up with it.

ML: When you were at school, were your teachers understanding?

MR: It was mixed, really. Some teachers were very supportive. There was one teacher when I was at primary school: she wouldn't patronise me, but she would sympathise. A lot of teachers were less sympathetic. But they had lots of other kids to deal with. I remember my dad picking me up from school at lunchtimes just to make sure I'd eat properly, because nobody would watch what I was eating. I couldn't eat much, so it was quite difficult.

ML: Did that mean you got left out socially?

MR: I think so, because I couldn't do things like playtime. When the kids went out to play, I'd stay in. I'd get to choose kids that could stay in with me, but I still missed out on quite a lot of interaction. And because I spent so much time in hospital, the people I used to hang out with would be the nurses. Hospitals have a nurse, or they used to, who was specially assigned to play with the kids. I guess I'd interact with her more than other children.

ML: Do you think that was also because you grew up quicker because of your experiences?

MR: I guess so. I think I was cynical as a child, which is quite sad, really. I didn't do the whole running round, falling over thing that a child does. When I was older, I was still cautious. You know when you lose that lack of inhibition when you're older? I don't think I even had it when I was little, because there was no way I'd fall over. There was no way I'd get any chance to do it. I never did things like that, like completely letting go and going bonkers. So I guess maybe I grew up a bit quicker.

ML: When you started going to the hospital on your own, did you find that quite a smooth transition?

MR: It was OK. Even when I was little, I used to ask questions for myself. And now I ask a lot of questions. I was a lot less stressed when I didn't think about my future. But I started asking things like, 'Will I be able to go abroad?', 'What about children?', 'Will I be able to have a job?' I guess these are the issues that are a bit more complicated for transplant patients.

ML: What was it like when you left home? Was that quite scary?

MR: Well, it was the first time I'd been by myself, so that took some getting used to. Since my transplant – up until last year – medically, I've been very low maintenance, except for my six-monthly check up. It's only with all these side-effects that I've had to keep going back to the doctors. So it wasn't really an issue.

ML: So you don't live your life in a perpetual state of anxiety?

MR: [Laughs] I do, but not because of my health – just in terms of life, you know, studying and career.

ML: So your condition hasn't in any way stopped you from having a normal life. Or are there areas where you have to hold back?

MR: Not really. I worked hard to get back to my education, and I try to do all the same things that everyone else does. Apart from this year, I don't think I've really had to hold back that much. Sometimes people say things like, 'Remember you're a transplant patient.

Remember to take it easy.' I find it very difficult to do that, because I try to keep up with everyone else. And this is the lifestyle in a big city – you're just trying to do lots of things at once.

ML: Presumably you have to be careful not to drink too much? And have you ever been tempted to smoke, like other young people?

MR: My lifestyle is very healthy. I've never wanted to smoke. I eat quite well. I drink lots of water, take lots of exercise, because I want to stay healthy. My biggest fear is being ill and not being independent, and having people look after me. So I try to take the best care I can. But that doesn't stop me going out and enjoying myself.

ML: How do you exercise now? Do you go to the gym?

MR: Yeah, I do! I really enjoy exercising. I didn't for quite a while, when I had gout. That was one of the most frustrating things – not being able to exercise and be up and about.

ML: Do you think your experience has enhanced your life in any way?

MR: Yes, I think it has. I guess I appreciate things a lot more. And I don't like to give myself excuses. I feel that now I'm better, I can do just as much as I want to – and I really want to. When I'm ill – it's probably a good thing and a bad thing – I just carry on, unless I'm really, really ill. Because I'm thinking, 'This isn't being ill. I know what being ill is like', and I just get on with things. I think that's one of the biggest things I had to learn: to slow down. You'd think it would be the opposite. You'd think, I'd have to try to push myself more, but it's the other way round for me. I've really had to learn to slow down and start putting my health first. Also, I think I'm incredibly lucky. And it's had an effect on what I want to do in my career.

ML: What are you studying at the moment?

MR: I'm doing an MA in Curating and Contemporary Design. My thesis is on whether or not hospital environments can affect your health. So I've kind of gone full circle. I guess it's to do with my experience when I was little.

ML: It must have been depressing for you going to one of those run-down, grotty hospitals when you were little.

MR: I guess it was. But I think with children's wards they always try a bit harder. I remember the playing ward being quite nice. People always make more effort for kids. I remember as a child visiting some of the adult wards, if I needed to see a specific person or get a test done. That was very depressing. I was very scared then. It was really, really grim.

ML: What sort of career do you see yourself going on to?

MR: I'd like to work in a hospital. It sounds really strange but I really like hospitals! They're places with so much potential and there's a feeling of community that you don't get with a lot of other places, especially in London. And I trust doctors – a lot.

ML: They saved your life.

MR: Yeah. They're my family as well.

ML: If you could go back and change a single thing that would have made your experience easier, what would it be?

MR: I wish I hadn't been so cautious after my transplant. I think that was due to other people telling me to be careful, but also because I'd been in this little box for such a long time. I was very cautious about everything: about sport, about what I could do and what I couldn't do. It's only since I've been at uni that I've really started exercising and taking more risks. I began to learn how to swim last year, and to do those sorts of things that I didn't do when I was little, when there was so much emphasis on getting back up to my educational standard. I remember trying to ride a bike after my transplant and being very, very scared. I missed a whole part of that childhood thing – you know, that sense that falling over is OK. I wish I'd done those things when I was little. But it doesn't matter. Whatever I've missed, I can do now.

The Beating Heart
Same Song – Different Rhythm

Ted Bianco

Some of the most poignant insights on life come from unexpected places. Who was not struck by the first images beamed back by the Apollo astronauts of the Earth hanging in the void with its precious cargo of life? Laid out before us for voyeuristic reflection was the biosphere we share with all the diverse life-forms of this planet. Thirty years later, the voyage of discovery was closer to home: it was the exploration through DNA-sequencing of the genetic blueprint of life. Written in our own genetic code, more clearly even than Darwin could have imagined, were the signatures of our common ancestry, linking us back through millions of years to the simplest single-celled organisms. The threads that bind us are powerful and emotive. One such thread is the beat of our hearts.

Life on this planet beats out a tune, but while the notes are the same, the tempo varies both within the life of an individual and amongst individuals from across the evolutionary divide. From his laboratory in James Cook University in Queensland, physiologist Geoffrey Dobson made this interesting observation as he pondered the subject of heart design and energy kinetics across the animal kingdom in 2003: whether you are incarnated as a shrew or a whale, you have a quota of around one billion heart beats. Use them fast or use them slow: that is the prerogative of each genus or species, but mind *how* you use them, because this is likely to govern your fundamental experience of life.

Size Matters – Or Does It?

Weighing in at 2 grams, the Etruscan Pygmy Shrew is one of the lightest amongst mammals, while the Blue Whale tips the scales at an alarming 100,000 kilograms. Their hearts vary in size by seven orders of magnitude – that is, ten million fold, or the difference between a drop of engine oil and a Volkswagen Beetle. Yet, remarkably, over the lifetime of these animals, each heart pumps a staggering 200 million litres of blood per kilogram of body weight. The tiny heart of the shrew delivers oxygen to the muscles, organs and brain just as effectively as the colossus beating in the chest of the whale. Before its working life is over, each heart will have fed 38 litres of oxygen to each gram of tissue in the body. Such consistency across the animal kingdom is testament to the underlying biochemical and biophysical processes that define the limits of our durability. We have in-built obsolescence, rather like the automobiles constructed for the mass market during the boom years of the 1960s.

But while the shrew and the whale may share some surprising statistics, they depart from one another dramatically when one considers the tempo of their lives. The heart of the shrew beats at an exhausting 835 beats per minute, while that of the whale pumps away at just 20 per minute. The shrew is dead within a year, having been profligate with its lifetime ration of heartbeats. The whale chugs along for the best part of a century. However, the high octane life style of the shrew does have a purpose: it needs to pump blood as quickly as possible to serve its super-charged metabolism. There may have been some laid-back shrews out there once, but when you're that low in the food chain, living fast and furious is clearly the strategy of choice to ensure that you leave your mark, including your genes, before succumbing to the pressures of predation.

The record for lethargy goes to an altogether different class of organism: this accolade belongs to the clam. At rest, a clam's heart beats just 2 times a minute. And even when excited, it only makes it up to a frequency of 20. At the other end of the spectrum, the heart of the Etruscan Shrew has been clocked at a record-breaking high of 1,511 beats a minute. It's hard to envisage a pump with so many moving parts working at such a rate. This is not a record that one should aim to emulate.

Pace Yourself

My heart rate is around 70 beats a minute, at rest, which is pretty typical for a member of my species. If I were fitter, it might be lower, perhaps as low as 55 to 60. When I run for the 18.54 out of Euston, it goes up to 130 beats a minute and with a bit more effort, I could safely push myself to 167: this is my Maximum Heart Rate, calculated according to the Londeree and

1 Pygmy White-toothed Shrew, smallest terrestrial mammal native to southern Europe, south-west Asia and north Africa

2 Immature sperm whale's heart, height approx 1 m

Moeschberger formula, 206.3 – (0.711 × Age). But when I was in my mother's uterus and just seven weeks old, my heart rate was even higher: then it peaked at around 185 beats a minute. That's just about the extent of the range of my pulse: a pretty modest three-fold spread from 60–185.

Even those who lie at the extreme ends of the spectrum for heart rate don't span a range much greater than this. But can we do anything to influence our heart beats? Doesn't yoga, for instance, allow us to reach a level of inner tranquility that transcends normal experience and calm the very beating of our hearts? Neurophysiologist Shirley Telles and colleagues from the SVYASA Deemed University in South India were similarly curious. But they were keen to go beyond mere speculation, so in 2004 they tested

3 Trials demonstrate that yoga reduces resting heart rates by an average 11 beats per minute

the hypothesis in a carefully controlled trial. They measured heart rate in a group of novices at the inception of their yoga training and again thirty days later. Pulse measurements were taken at rest and repeated after six minutes of yoga, during which the students focused on depressing their heart rates. The same measurements were taken from a group of volunteers who did not receive training. The results were intriguing. Before training, none of the volunteers could lower their pulse significantly by conscious effort. But following training, the yoga students shaved an impressive 7 beats a minute off their heart rates when they entered a state of karma. And their resting heart rates were reduced by an even larger margin – by an average 11 beats a minute. In contrast, the untrained group had similar readings throughout the experiment.

So yoga's reputation appears to be well founded – it does seem to equip individuals with the ability, albeit limited, to control their heart rates and add a modest 10 percent to the range of their pulse. But what about the other end of the spectrum – what elicits a full-on response? Well, we can all recall those occasions that got our hearts racing – opening the mail to discover whether you got that job, standing up to give the dreaded after-dinner speech, plucking up the courage to ask the pretty girl with the pony-tail for a date. At moments like this, we know our hearts are pounding, but who among us has had the fortitude to measure their pulse – let alone record the measurements with the precision demanded by science? Time to introduce John Deering. Mr Deering had such fortitude; indeed, he left us with one of the most gripping records of the response to acute anxiety we have. Not that Mr Deering was a paragon of virtue. After all, he was a convicted murderer, who ended his days on death row in Utah State Penitentiary. By the time of his death in 1938, he had resolved to make good, and he offered his mortal remains in the service of science. He donated his body to the University of Utah and his eyes to a grateful specialist in California. But of greatest interest to us here was that he offered his dying moments to the local cardiologist, who enthusiastically wired him up to an electrocardiograph device just before he was led into the room where he was to be shot. The hapless Mr Deering saw his heart rate rocket from 72 to 180 beats a minute at the moment he was strapped down in front of four loaded carbines. His heart stopped 15.6 seconds after he was struck by the bullets. Now that's what I call an unrepeatable experiment.

If humans had the propensity to control heart rate with ease, there would be no place for a game show like 'The Chair'. Premiered by the ABC network in the USA in the fall of 2001, it had a character all of its own. The show made a brief appearance in the UK, but was probably too biomedical to appeal to an audience reared on a diet of 'Blind Date' and the 'Wheel of Fortune'. The game comprised some familiar ingredients: a series of questions to the contestant, prize money that went up exponentially to heighten the tension, and a chair sitting in a bright pool of light for dramatic effect. But that's where the familiarity ends. The core concept was that players were required to maintain their pulse below a certain threshold while under the onslaught of unremitting pressure. The threshold was set at 160 percent of the contestant's resting heart rate. So, for me, this would be just 112 beats a minute. That's the kind of reaction I experience on seeing trout rise around my fly. Little chance, then, that I'd come away from this show any better off. And just to make life difficult, the participants were faced with 'heart-stopper' events at intervals during the competition: for example, having a close encounter with an alligator, or lying in the path of a tennis ball as it was served at great speed past their heads. In a sophisticated nuance of the rules, players were not simply disqualified for exceeding their maximum heart rate. Instead, they lost money for every second over the limit. Cruel indeed: imagine trying to calm down when you know that every second is costing you $1,000. Over the life of the show, only one contestant won the maximum $250,000 and only three got to the last question. And to further illustrate the difficulty of controlling your heart rate, one player who made it to the last question managed to lose $132,000 worth of prize money by failing to get his heart rate under control for just 2 minutes 12 seconds. From an evolutionary standpoint, when it comes to getting a grip on the behaviour of one's heart, human beings are simply not that advanced.

Timing Is Everything

So, how is heart-rate regulated and what fulfils the all-important role of the metronome that sets the beat within us?

The heart's natural pacemaker is called the sinoatrial node. This is a small mass of cells overlying the right atrium. Like the god Janus, it has a dual outlook on the world. That is, it exhibits the properties of both muscle and nervous tissue. When the node contracts, wearing the face of a muscle, it generates an electrical impulse. Now it

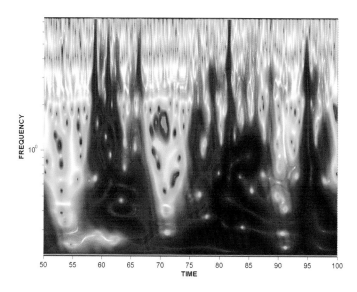

4 Computer-enhanced image from an electrocardiogram

behaves like a neuron, discharging a signal that travels through the cardiac tissues to the atria (or upper chambers) and down to a relay station called the atrioventricular node. Each electrical impulse from the sinoatrial node stimulates contraction of the atria and a short while later, contraction of the ventricles (or lower chambers) in a tightly orchestrated event. The trick is in the interplay of the sinoatrial and atrio-ventricular nodes. The sinoatrial node sets the beat, and the atrioventricular node delays passage of the signal to the ventricles so that the atria have time to contract fully. Once the atria have pushed blood into the ventricles, it's time for them, in turn, to contract. But what informs the sinoatrial node about the body's changing demand for the circulation of oxygen? It is the autonomic nervous system – that part of us that operates without conscious thought. As with most things that work well, we go through our lives blissfully unaware of what needs to function effectively in order to keep us going. But with every impulse sent out by the sinoatrial node, one of the most fundamental life-support systems we have quietly gets on with its job.

Cheat If You Have To

'If it can go wrong, it will go wrong': so reads the age-old adage and, sadly, this applies on occasions to the machinery that makes our hearts tick. Well, if one cannot succeed without a bit of help, so be it. Enter Wilson Greatbatch, World War II veteran, aviation radioman, electrical engineer, and holder of no less than 140 patents. Impressive enough, but that's not why he'll be remembered. His place in history is secured as the inventor of the first implantable cardiac pacemaker. This is an invention that is credited

5 Wilson Greatbach

with having saved millions of lives across the world since its creation in the 1950s. In a classic example of the rocky road to discovery, Greatbatch's claim to fame arose from an accident – indeed, from a mistake he made while building an oscillator to record sounds from the heart. He inadvertently fitted a resistor of the wrong specification, and it began to give off a steady electrical pulse. He was quick to see the potential of the device to regulate heart beat. An earlier apparatus from John Hopps had demonstrated the principle, but was clinically impractical. That instrument was the size of a television and was painful to use. Greatbatch gave up his day-job and repaired to the barn he used as a workshop, where he threw himself into the task of refining the concept. Time was against him – he had just $2,000 on which to live and support his family. But his resolve was unshakeable. Within two years, he had created the world's first implantable pacemaker, and a little while later corrosion-free lithium batteries with which to power it. Should the spark ever go out of your life, or more specifically your heart, it's comforting to know that there's always Wilson Greatbatch's clever little invention to fall back on.

Masters of control

Whereas *Homo sapiens* is no expert in the regulation of heart rate, there are some real aficionados out there. Perhaps the most adept is the hummingbird. Clearly, this is an extremely active creature, with a high oxygen demand dictated by the niche it occupies: essentially, it is a bird that behaves like a bee, only it's quite a bit heavier. When it hovers in flight to collect nectar from flowers, its wings are required to work at around 50–200 beats per second. Even level flight is conducted at exhausting speed – when humming-birds flit from tree to tree, their 'cruising speed' never drops below 35 miles per hour.

To fuel such performance is energy intensive, and this means that the heart must work hard to deliver sufficient oxygen from the lungs to the flight muscles. So, while the birds are feeding, the heart pumps at something like 1,000 beats a minute. When resting between flights, it drops to around half the in-flight rate. Hummingbirds fly by sight, so they are active only by day. At night they find a roost, but they don't simply go to sleep. They enter a profound state of torpor. The body temperature drops from 40 to 21 degrees centigrade and the heart slows to a mere 50 beats a minute. Why is this and what does it mean in terms of the life of the hummingbird? It all boils down to the bird's chief characteristic – the manner of its feeding. Nectar is essentially sugar, which is a great source of immediate energy, but one that's quickly used up. To remain fully active, hummingbirds must feed every fifteen minutes. They do, of course, acquire protein in the form of the occasional insect, but this doesn't constitute the mainstay of their diet. What they need is an energy reserve to make it through the night. But they store only meagre quantities of glycogen (a complex sugar laid down in the liver and muscles as an energy reserve) because it is bulky, and they try to hold on to their

6 Hummingbird

equally modest stashes of fat for their seasonal migrations. So what is a sleepy hummingbird to do? The conundrum was settled by the evolutionary adaptation of shutting down the body's metabolism. And the sweetest thing of all is that the heart is spared from all that pumping during the bird's slumbers through the night. It may be more than coincidence that hummingbirds live surprisingly long for such frenetic creatures. Typically, they survive from four to twelve years, depending on the species, many times the life-span of a shrew. Since we, ourselves, spend a third of our lives in bed, it makes you wonder about all those 'wasted' heart beats.

Sleeping It Off

There are animals that habitually indulge in far more serious slumbers than humming-birds – I refer, of course, to creatures that hibernate. The term 'hibernation' is used loosely in everyday parlance, but to the biologist it has a quite precise meaning. True hibernation is a specialised form of adaptive, seasonal hypothermia. The authentic hibernator undergoes a major slowing in metabolism, a profound reduction in core body temperature and a depression in heart rate. Many reptiles and amphibians of northern hemispheres hibernate. Some, like the Wood Frog and Painted Turtle can even withstand partial freezing. In the icy conditions of mid-winter, the heart rate of some freshwater turtles may slow to a mere one beat in ten minutes from a high of 40 beats a minute on a warm afternoon in July. Mammals such as woodchucks and hedge-hogs are also true hibernators. For example, body temperature in the European Hedgehog can drop by 25 degrees centigrade and heart rate fall from 190 to 20 beats a minute. Energy consumption drops to just 5 percent of normal. This is profound metabolic shut-down. Many mammals, like Grey Squirrels and bears, however, simply enter a deep state of lethargy. This is not totally unlike hibernation, but is distinguished by less profound torpor, frequent arousal and periods of normal activity. Predictably, there are debates among naturalists about certain species like Black Bears, which maintain a relatively high body temperature yet are considered by some to be true hibernators. Their heart rate may fall to just 8 beats a minute. But beware. In a foolhardy experiment, one naturalist decided to study the rate of arousal of a Black Bear from its state of hibernation. Showing the true grit of a dedicated biologist, Lynn Roberts clambered on hands and knees into the den. He pressed his ear to the chest of a five-year-old female but could hear nothing. He concluded that the heartbeat was too faint to detect. With gentle prodding, he picked up a discernible pulse. A few moments later, the bear lifted its head. Time to beat a hasty retreat. In less than five minutes, its pulse had soared to 175 beats a minute – higher than would be expected even for a bear firing on all cylinders.

Behind this anecdote is an interesting phenomenon. During deep hibernation, the autonomic nervous system is essentially turned off. When it kicks in during arousal, it is the so-called sympathetic arm of the autonomic system that swings into action. This

Der Bæhr im Lager an den Bratzen Saugent.
N.72.

7 Johann Elias Ridinger (Germany, 1698–1767), *A Bear Resting in its Cave*, etching with engraving, 18.5 × 15 cm

is the system responsible for arming the body in 'fight or flight' situations: the system that makes the hairs on my neck stand up when I greet someone I'm trying to impress by the wrong name. So Mr Roberts was wise to make an undignified exit from the bear's den. This mechanism has obvious utility. On waking, the bear must quickly restore its body temperature and other essential functions. So the heart goes into override, rather like the engine in the old banger I drive when I use the choke to start it in January.

2001: A Space Odyssey And All That

In years to come, human beings will reach new galaxies by passing the years in a state of suspended animation within their little silver crafts. Or so the movie-makers would have it. The only cloud on the horizon is that science hasn't quite caught up. Suspended animation, while practised year after year by toads and terrapins, is uncharted waters when it comes to humans. Still, the signs are that it's getting a little closer. Witness the use of whole-body cooling to shut down the circulatory system during cardiac surgery in Russia. Or the reports that the stuff of schoolboy practical jokes, hydrogen sulphide (which gives stink bombs their rotten-egg stench) induces a dramatic torpor in mice – importantly depressing heart rate, breathing and body temperature without an accompanying collapse in blood pressure. But there remains a long way to go.

The Threads That Bind Us

At just twenty-two days after conception, you and I were little more than tadpoles. But we were tadpoles with hearts, and ones that were already working hard. We were not alone. For every animal with a circulatory system puts in the plumbing as one of the first priorities of embryonic development. Not only that, but we all employ the same basic genetic toolkit to achieve this evolutionary triumph. In the middle of the action is a gene that shares its name with a central character in the Hollywood classic, *The Wizard of Oz*. The gene is called *tinman* after the loveless comrade of Dorothy, who bemoaned the fact that he had no heart. This discovery came from the weird and wonderful world of fruitfly genetics. Those in this field are well known for their infantile tendencies – naming the genes they discover after all sorts of nonsense. They have genes like *sunday driver* (named after a mutation that leads molecules to get lost as they journey through nerve cells); *ken and barbie* (christened after a mutation in which the external genitalia are missing); and *british rail* (a gene that suppresses the expression of another gene called *always early*). Perhaps geekiest of all is *cheap date* (mutants of which are especially sensitive to alcohol).

In spite of my derision, the name *tinman* is surprisingly apt, because it was applied to a gene found to be damaged in mutants that fail to develop a heart. This is such a serious mutation that the flies cannot complete development. They die within the egg. It's what's known in the trade as an 'embryonic lethal'. *Tinman* is an ancient gene found in us too (although under the rather dry name of NKX2.5), and performs a critical function so fundamental that it has been passed on through evolution for hundreds of millions of years. It was already an old favourite at the time of the dinosaurs. Indeed, it has been traced back 700 million years to a common ancestor that was a flatworm. And why is this? Well, its job is to control the function of other genes – it's a so-called 'homeodomain transcription factor', which acts as a master switch during the programme of development required to build a viable individual. By orchestrating the cascade of gene expression that creates the components and assembly line required to construct a heart, it lies at the top of a command structure. Take away the general, and

8 Jack Haley as the Tin Man, in Victor Fleming, *The Wizard of Oz*, 1939, film still

the soldiers are impotent to execute the battle plan. *Tinman* literally gives us a heart – as it does to each and every organism whose breast beats to the tune of this enduring organ. The threads that bind us are powerful indeed.

For Barbara

Gurleen Sharland

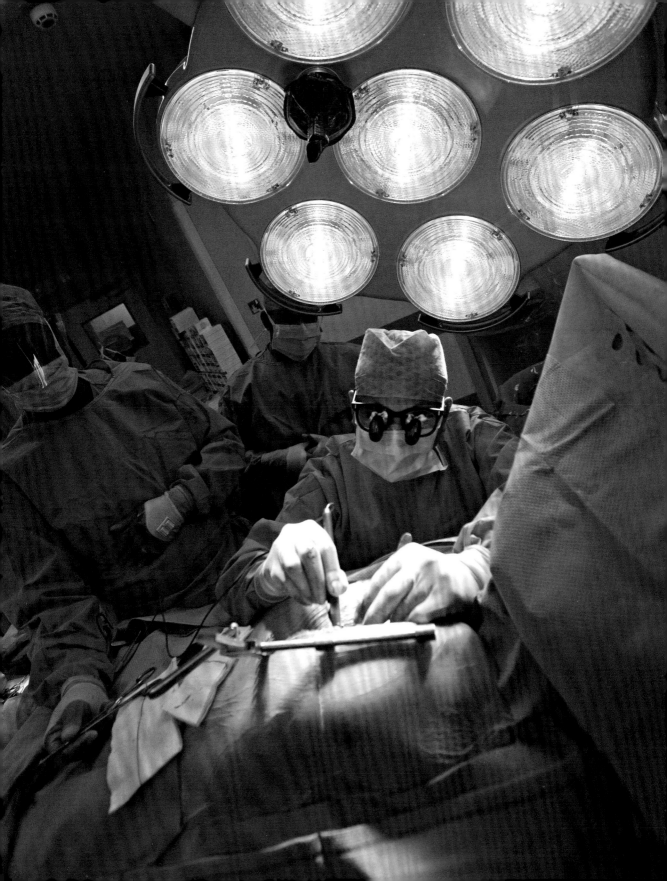

Gurleen Sharland

Consultant Foetal Cardiologist, Evelina Children's Hospital at St Thomas's, London, in conversation with Melissa Larner

MELISSA LARNER: Can we start by talking about women in cardiology? There seem to be very few of you, especially female heart surgeons.

GURLEEN SHARLAND: There are a few female surgeons. There's one at Great Ormond Street, who does children's hearts. And I'm trying to think who else . . . But even if we're just talking about cardiology, and not surgeons, it's still very male-dominated.

ML: Why is that?

GS: I think it's the way medicine has evolved; it's not very child-friendly if you have a family.

ML: Because there's a lot of operating in the evenings and at night?

GS: Yes, and it's quite . . . 'high powered' is the wrong term, really. It's not that it's more high powered than the other specialities, but it's *perceived* as being a speciality for certain types of personalities.

ML: Yes, Martin Elliott at Great Ormond Street was talking about this – the old, macho, 'cowboy' image, which is now being tempered by teamwork.

GS: That's certainly changed for the better. Now, it's very much a team thing – not one person who does everything. Certainly here, at this hospital, it's a team approach. It's all one big team, and all the team members speak to each other about what we're planning, and it includes not just the cardiology teams and the surgeons but the obstetric team, the doctors from the neonatal unit and the intensive-care units. We all have to know what's going on.

Pradeep Narayan, Registrar to Professor Gianni Angelini at Bristol Royal Infirmary, opening up the sternum and chest cavity, 2005, digital photo by Mike St Maur Sheil

ML: So in fact, it's the perfect job for a woman – that kind of communication and team spirit.

GS: Yes, I personally don't see any reason why women shouldn't do it. But having said that, my colleagues are all men. [Laughs] We've recently had a woman appointed who deals with the adult population that have congenital heart disease, but it's going to take a while to change, because there's this taboo from the past. But I think women should be encouraged to do it; it's something they can do. There are certainly more female trainees than there were.

ML: What do you think would encourage more women to go down this path?

GS: I think to get away from the macho kind of image – the idea that you have to be that type of person; to recognise that there are different aspects to it. OK, some things might not suit, but there will be other parts of the service that do suit you, and you can be part of that team. For most women in medicine, though, it's all to do with whether you're going to have a family or not. That applies across the board and I think that women have chosen specialities that have allowed them to take time out easily and to come back on the ladder at a similar level. The problem with a lot of specialities is that technically you're allowed to have the time out, but you know that you're not going to come back on the same rung, whether you're going to come back full-time or part-time, and if you work part-time, despite what anyone says, you don't quite reach the same level. That shouldn't be the case with equal opportunities, but it's the reality, and women are aware of that.

ML: Have you found it a struggle to get to the top?

GS: It has been a struggle, I've got two children, but both times, when I had my babies, I had very little time off and if I hadn't done that I wouldn't have reached the level that I did reach. My first baby was twelve weeks old when I came back to work. Now I think, 'God, that was ludicrous!' I should have had several months off. I think there's a lot of issues like that. Nowadays, a lot of women want to take a year off. If I'd taken a year off, I don't think I'd be sitting here now, in this job. But it's getting easier for women, and hopefully with time, it'll become more the done thing to have a family, to take time out and to come back at the same level.

ML: When you began your career, did you feel you had a vocation in medicine?

GS: I think I did, yes. I've always known I wanted to do medicine.

ML: Was paediatrics always your ambition?

GS: No, I started with adult medicine, actually. I did some adult medicine first, but I was always attracted to paediatrics. And once I did paediatrics, I don't think there was any doubt that I wanted to do anything else. And then I came into the field of cardiology within that. I'd always been interested in cardiology.

ML: What was it about the heart that drew you?

GS: I was just always fascinated by the workings of the heart. And to me, that's the centre of life. You've got to have your heart to live. I know the brain's important, but if you haven't got a heart, you're not going to live.

ML: Could you explain what you do? Your speciality is the diagnostic side, isn't it?

GS: Yes. We do heart-scanning in foetuses. The bulk of our work is around diagnosis of heart problems in the womb and then planning the treatment after birth. I've been doing heart-scanning in foetuses since 1987, and have been a consultant since 1993.

ML: So you must have been working in this area nearly from the start. For how long has this particular branch of technology been around?

GS: It was started in the early 1980s, so I suppose I was in there quite early on.

ML: You must be one of the most experienced in your field.

GS: Probably, yes. The first unit was started in 1980/1, at Guy's hospital and that was one of the first in the world, and the centre there is still one of the most renowned in the world. Now there are more centres around the UK that do foetal scanning, but not many are dedicated just to the foetus and heart disease, as we are here at St Thomas's. So that's what makes us unique.

ML: Presumably the technology is improving all the time?

GS: Oh yes! If I showed you images from when I first started compared to what we can get now, you'd really see the difference. At that time, we used to do the scans at twenty weeks. Now, we can do them from fourteen weeks. The technology's got better, so the resolution is much clearer. Obviously, there are lots of factors that influence it – with a slim lady you get much clearer images then you do with a larger lady. The baby's position also affects it: if they curl themselves up or hide away then that makes it more difficult. But on the whole, nowadays, you get some very nice, clear images of the heart.

ML: And are you normally just diagnosing conditions, or do you treat them at this stage?

GS: There are a few cases where you'd consider treatment before birth. These fall into two categories. They're either babies with a disturbance of the heart rhythm, and we can treat some of those by giving the mother drugs that cross the placenta and control the baby's heart rhythm, or they're structural problems, where there's an obstruction within the heart due to a blocked valve. In this situation, we can use balloon catheters to try and open up the area of blockage.

ML: That's extraordinary. How do you get the catheter into the baby's heart?

GS: You have to go through the mother's abdomen, straight through the baby's chest, and directly into the heart. So you're very dependent on the baby's position. The last time we did a procedure, we were waiting for hours for the baby to get itself into a position where we could access the heart, because it was hidden away. We had the whole team of about twelve people just hanging around waiting for this baby to move! [Laughs]

ML: And it was a foetus of how many weeks?

GS: That one was twenty-four weeks and had an obstruction to the valve leading to the lung artery, so we were trying to open that up.

ML: Will that actually cure the baby?

GS: No, it's really a holding thing. With most congenital heart disease, the heart's formed like that and you can't cure it. After birth, the surgeons can 'correct' some types of heart disease – they can do repairs. And there are some forms that we can't correct. Then what we're doing is palliation – trying to make the blood reach the right place – but we're not actually correcting the malformation. The interventions that we do in the womb are to try and open up obstructions, to try and make the affected side of the heart work the best we can. This is because if a ventricle is pumping against an obstruction, it may stop growing. By relieving the obstruction, the idea is to give the ventricle a chance to develop more normally than otherwise, so that the result of surgery afterwards is likely to be better than if you hadn't done it. But it's not the definitive operation. It's the first step in whatever else is going to be needed.

Having said that, it's not something you do every day; those are select cases. With the vast majority of babies, it's more about diagnosing the problem, informing the parents about what the problem is, and the implications, and then you can begin planning. If the diagnosis is made when the baby's still in the womb, it gives parents time to prepare. Then they can come back and ask questions, we can put them in touch with other parents who've had a similar problem, we can let them speak to our surgeons if they find that helpful. We can think about looking at the rest of the baby. Is there anything else wrong? Do they need to have an amniocentisis? Can you imagine what a shock it is to be told that your baby has a heart problem? Especially as we're talking about major problems; we're not talking about small holes. However, for them to try and digest all this information for the first time after birth, while the baby may be being whisked off to surgery, is probably worse. Antenatal diagnosis gives the parents a chance to prepare for what's coming, so that by the time the baby's born, they know what's going to happen, they know they need an operation and they already know what the intensive care looks like because we've shown them around.

ML: Does everybody routinely have a scan to check their baby's heart?

GS: No, what should happen is that the majority of pregnant women – at least 90 percent of women in this country – have what's called an 'anomaly scan', where obstetric sonographers look at the whole baby. And what they should be doing during that scan is examining a few views of the heart to make sure it looks OK. The detail in which they do it isn't the same as we'd do here, but we've taught them a few screening views, and if these don't look right, or they can't see the views, or they're a bit worried, then they refer those ladies here. There isn't widespread enough expertise nationwide to allow every single woman to have a detailed heart scan, so we're dependent on people in obstetric units doing a basic screening and then saying, 'Hang on this doesn't look right', and then sending them to more experienced centres like ours.

ML: And the only way they'd really know would be through a heart scan – there wouldn't be any other indicators?

GS: No, there wouldn't. But there are some women who are at high risk. A lot of the women we scan here are at high risk, most commonly because of a family history. If they've had a previous child with heart disease, or they've lost a baby because of a heart problem, we'll offer to scan them in their next pregnancy at fourteen weeks. If one of the parents themselves has some form of heart disease, then they're also at an increased risk of having a baby with a heart problem. And diabetic women who take insulin are at risk, or women on medicines for epilepsy. A new test performed nowadays is measuring the neck thickness in babies at around twelve weeks into the pregnancy. This is done mainly for Down's Syndrome. But in fact, we've found a large number of babies with abnormal neck thickness who don't have Down's, but do have a heart problem, so that group gets referred for heart scans. Although we scan a lot of high-risk women, most of the problems are detected through screening the low-risk population during obstetric scans. Over 50 percent of those referred from obstetric screening will be abnormal. If they've had the neck-thickness measurement at twelve weeks and it's abnormally high we can see them at fourteen weeks. But a lot of women don't get referred till later, so the average time of a scan is around twenty weeks, and that's because of the way they get referred, not because we can't do them earlier than that. And for certain groups of people, we'd also want to see them in later pregnancy, because there are some types of heart problems that might be progressive – it might not look so bad at sixteen or eighteen weeks, but it could get worse by twenty-eight weeks. That group's important, because that's a group in which we'd consider doing an intervention in the womb to stop it getting even worse.

ML: Can you describe some of the different types of problems that you're dealing with?

GS: There's a difference between the types of heart disease babies get and the types of heart disease adults get: what we're talking about is very different from coronary artery disease or something like that. Essentially, what we're seeing is problems where the heart hasn't formed properly. You can have a chamber missing, you can have valves missing, vessels that haven't formed properly, vessels that come out the wrong way round. You get all combinations of things. A lot of what we see is very complicated anatomy. Having a hole is one of the more simple forms – although it might be a more major problem, it's simpler to fix.

ML: How is a hole in the heart dealt with?

GS: In simple terms, it's patched up. It's like, if you've got hole in your trousers, you apply a patch with stitches. So a hole in the heart is actually one of the least serious things. The types of problems we're seeing are much more complicated than that. A lot of the types of problems we see, can't be fixed back to normal. There's huge variation, but what we're seeing most of the time is the more complex end. We sometimes see the more simple stuff, which is nicer, because you can be more positive, but the spectrum we're seeing is skewed towards the more severe end.

ML: Is the success rate good?

GS: It depends on what it is. Surgery's got a lot better. Since I became a consultant, surgical techniques have improved a lot. So what I say to parents now is different from what I was saying to them when I began. But there are still some things that you just can't do anything about. Or even when you've done something, the outlook isn't going to be good. It's just that the child has a bad heart and you can't do anything about it. And there are some conditions now where surgical techniques have allowed the child to survive, where previously they wouldn't, but what we don't know is for how long they're going to survive. If the parents say, 'Is my child going to have a normal quality of life', the answer has to be 'No', in some of these complex cases. But for some of them, we're waiting for these kids to grow up to see what happens. We know some of them can do OK in the short term, but we're anticipating problems in the long run.

ML: They might have to have more operations?

GS: Yes, or they might have to have a heart transplant when they're grown up. With all that in the equation when their baby's not even born, it's quite hard for them to take it all in. So the information we give to each parent has to be individualised for them and their baby depending on what we're seeing on the scan. For some types of heart problem, you can say, 'Well, once we've corrected this, your child should be able to do most of the things that other children do', but for some of them you know that's not the case. Even the ones that have been corrected, you still have to keep an eye on them and you still have to bring them in for review. It might just be once a year, but they still have to come back to make sure things are OK, and you don't know if something might go wrong as they're growing up. Parents always have that in the back of their mind. So it's not usually a one-off thing where we can say, 'Don't worry, we'll fix it and that's it.' It's a lifetime commitment.

ML: Does being on the diagnostic side protect you a little bit from the trauma if a patient dies, since they're no longer directly in your care?

GS: No. You've known them since before they were born. You've looked after them, and you know their families. And every patient who dies, even if I don't know them, does matter to me, because it's going to affect what I say to the next set of parents where the baby has a similar heart problem.

ML: Do you find it difficult to deal with the highs and lows?

GS: Yes, and we certainly have them in the unit as a whole. We have high moments and low moments and the low moments affect everybody: the nurses and technicians working here as well as the doctors. Sometimes the babies may die even while the woman's pregnant, or they may die soon after birth, when your link with them is closer than if they'd died when they were a few years old. Because I've worked in the same unit for a while, there are some patients whom I saw as babies, and who get presented here as teenagers now, and I think, 'Oh yes, I remember them.' So even though I don't personally look after them, I'm very much aware of it when something happens to them. And sometimes I bump into parents and they still remember me. It's a very emotionally challenging job. It's not an easy job to give people bad news. What I always say to my colleagues is, 'By the time they get to *you*, the parents know everything, so you haven't got that same emotional challenge.' But when they come to see us here, they're expecting a normal baby, and to destroy that dream and to say, 'Well, no, you haven't got a normal baby, it's got a serious heart defect' – that just destroys everything, and to help them to cope is very demanding. I don't think you can completely detach yourself from that. I don't think you'd be human if you did that. Even just the giving of bad news to somebody is quite hard, and I think you need to have an outlet for it. I'm quite fortunate: my husband is also a paediatrician and I can bounce things off him and I know he understands how I feel, but you don't want to take everything home with you.

ML: Especially if he's had a bad day too.

GS: Well, it does occasionally happen. You just sit in the kitchen and get a bottle of wine out. [Laughs] But it does help to have someone who can empathise, and sometimes you do need to vent your feelings because they've got to go somewhere. But I'm wary of taking it home too much because I don't want the children to be affected by what I do at work.

Another difficult thing that we haven't talked about is that when women present before twenty-four weeks, and there's a major problem, they have the option of termination, and that's one of the discussions that comes up with a significant number. These days, that's not something the majority choose: the termination rate is 29 percent at the moment. When I started, it was 70 percent, so it's really come down – because the treatments we can offer have improved and the success rate of the treatments is better. Society has also changed. The way women view termination is different. A lot of mothers are having their babies later, so they may have fewer chances. I don't think that's the main factor, but it could be a factor. What's also happened in this particular unit – and this isn't true of everywhere – is that we're seeing more babies with an isolated heart problem. When I first became a consultant, when we found a heart problem, it was quite common that they had a chromosomal abnormality or some other abnormality and obviously that sways the prognosis, which will sway the parents' decision. But now we have much more isolated heart cases, because the others get filtered out with the neck-thickness screening. So, if you looked at the numbers over the last five years, we'd probably find that only 10 percent have a chromosomal abnormality, which is the same as what you find in the paediatric population, whereas it used to be 20 to 25 percent.

But I think parents' own views have changed in the way they approach this. It's all to do with quality-of-life issues. Some parents can't face the idea of losing a child in later childhood. The baby might survive for a few years and then die when it's six, seven, eight, nine, ten, and the prospect of that is difficult for some people to face. But views vary on this. Some people say, 'Well, whatever we can do for however long we have the child, that's fine.' But others say they just couldn't manage that. Some have other children, and they think about the effect on the other children of a sibling that's going to be in and out of hospital. There are so many factors involved in it.

ML: It must be a tough decision for you to advise on.

GS: Yes. Many parents will say, 'Well, what would you do?' And I say, 'The decision I'd make would be under different circumstances from yours and I can't tell you what I'd decide.' I've never, ever told a parent what I'd do. Most times, you're not actually sure what you'd do, even with all the knowledge. I usually just say to them, 'You have to make the right decisions for you as a family, and you have to make a decision you can live with. There's no absolute right or wrong decision. You have to make the right choice for you as an individual.' And sometimes there's conflict between the couple – one wants to do one thing, the other wants to do something else, and marriages can break up because of something like that.

ML: To look on the bright side, these are babies who probably would have died very early on in the past.

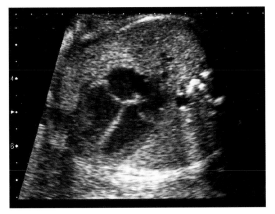

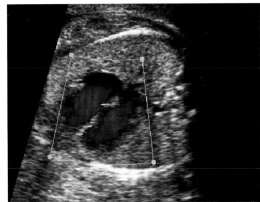

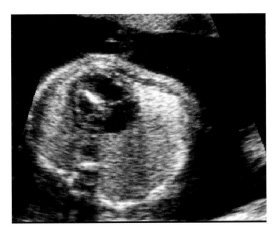

Foetal scans: **a** Normal 4-chamber view of the heart, showing that the chambers on both sides are equal in size. **b** Normal flow into both ventricular chambers, demonstrated using colour. **c** 4-chamber view in a form of hypoplastic left heart syndrome. The left ventricle is abnormally developed and the walls of the chamber appear unnaturally bright.

GS: Yes, certainly a group of these babies did. For example, fifteen years ago, when we diagnosed a baby with hypoplastic left-heart syndrome, where none of the left-heart structures have developed, I'd just say: 'There's nothing we can do for this condition.' In the womb, it doesn't matter because the two sides of the heart work together, but there's a structure between the two main arteries called a duct that shuts after birth and when that shuts, no baby with this condition can survive. Fifteen years ago, that's what I was having to say to parents – 'There's nothing we can do. Your options are to have a termination or to carry on with your pregnancy and let nature take its course.' And we put a few babies up for heart transplantation at that time, but in fact they all died waiting for a donor; we didn't have any successes, so I wasn't really pushing people in that direction. And then in America they developed new surgery called the Norwood Procedure. Bill Norwood was the first person to do it and there have been various modifications of the technique. In 1995 we asked Mr Norwood to come here, we were at Guy's Hospital at that time, and he taught one of our surgeons how to do the Norwood.

ML: What is that technique?

GS: When you have hypoplastic left-heart syndrome you've got one pumping chamber –
your left-side one hasn't developed and you've only got one big vessel. The Norwood
involves re-creating the vessel that comes out of the heart, so that it then pumps blood
into the body. At the same time, you create a shunt into the lung arteries to maintain
the lung circulation. And then at a later stage you redirect the blood that's coming back
into the heart from the body, so it goes straight into the lung circulation, by-passing the
heart. It's a variation of the Fontan circulation. So we started offering the Norwood
Procedure in 1995, and now about 50 percent of parents will carry on with their pregnancy following the diagnosis of hypoplastic left-heart syndrome. Currently we're
quoting 80 percent survival from the first stage of the Norwood operation, but the
longer-term prognosis remains guarded.

ML: Survival right into adult life?

GS: No, this is short-term survival. And these babies haven't grown up yet. The first one we
did was in 1995, so that one's only eleven. I think in Birmingham they've got a child
who's twelve or thirteen. In America, they've got some who are older, but then others
have died later. We can do wonderful things, but we don't know the implications of it
all for the long term, and for my job, that's very important, because it may be that in
another ten years, I'll have to start changing what information I give to parents about
their child's heart defect.

A lot of surgical techniques have changed in recent years. Transposition of the
great arteries – the arterial-switch operation – has changed things dramatically. I
have great admiration for my surgical colleagues. I think they're all fantastic and do
a great job with very difficult things, but there is a limit to what they can do. More
things may develop in the next five to ten years. We may be offering things that we're
not offering now.

ML: Do you give babies complete heart transplants?

GS: We don't usually offer it for a newborn baby, partly because there are no donor hearts.
In years gone by, there used to be donor hearts from babies that had anenchephaly –
where their brain hadn't developed, so they couldn't survive. But now, those babies all
get picked up on antenatal scans, so most of them are terminated. Also, the results of
the transplantation weren't good for newborn babies, so we thought it was better to go
down the Norwood line rather than the transplantation line. I think in North America
they still offer it, but we don't offer it as an option here. When children are older it could
be considered, but it still depends on various factors, one of them being donor-
availability. That's the sad side of it. Perhaps in years to come we'll have mechanical
hearts. That might be a development of something in the future that will change things
for people, but we're not there yet.

ML: If there was one change – technical or in other fields – that would improve people's chances of surviving problems with their heart, what would you want that to be?

GS: I'd like to know what causes heart disease. We don't know if it's caused by a gene or several genes. I'd like to be able to identify them and to do something about it – to stop it actually happening. It would be nice if in my lifetime, we could find what causes it, even if we don't get as far as actually preventing it, because I think that's the most frustrating thing for parents. They often ask, 'Why has this happened?' And I say, 'Nature got it wrong when this baby's heart was being formed.' Well nature *has* got it wrong, but why has it got it wrong? We usually say the cause is multifactorial but no one knows exactly what it is. It's likely that environmental factors influence the development of congenital cardiac abnormalities, but until we know what actually causes it, we can't stop it happening. So at the moment it's diagnosis and treatment, but we need to go backwards to prevention. That's what I'd like to see happening. Then, of course, I wouldn't have a job!

The Broken Heart
The Story of 'Heartbreak Hotel'

Michael Bracewell

Hot dog, Mae! Play it again!

> Elvis Presley to Mae Boren Axton, on hearing
> the demo tape of 'Heartbreak Hotel'

We're only given as much as the heart can endure.

> Patti Smith, 'Farewell Reel', 1996

Ever since the mid-1950s – when 'Heartbreak Hotel', the principal subject of this essay, was recorded – and up to the present day, pop music in all its myriad forms has retained a fixation with the heart as an image, metaphor and symbol. From Elvis Presley himself, through Lou Reed's deadpan homage to rock and roll as a life force, 'Rock and Roll Heart' (1976), to Pet Shop Boys' disco melancholia, 'Heart' (1987), or Blondie's UK Number One, 'Heart of Glass' (1979), the pop song has never tired of hymning the heart. (In 1982, Olivia Newton John would even have a hit with the unfortunately titled 'Heart Attack'.) The heart in popular song stands not just for love in its romantic and erotic sense, but for the vulnerability of the self and the inner self – the fragile, sacred core of one's being and the capacity to feel alive to the world.

Pop music can sing about the heart's estate with a contemporary, metaphorical wit that only a form that is based on the conflation of sex and technology (as pop is) could achieve. In terms of sheer lyrical cleverness, almost at a Cole Porter level of ingenuity, we could look to Stephin Merritt's extraordinary recording with The Magnetic Fields, 'Epitaph for My Heart' (1999). Merritt's epitaph begins:

'Caution: to prevent electric shock
do not remove cover.
No user-serviceable parts inside.
Refer servicing to qualified
service personnel.'
Let this be the epitaph for my heart

He then goes on to name-check one of the most venerated institutions in the history of popular music – the building in New York where, in the early 1960s, the very young songwriting team Carole King and Gerry Goffin wrote some of their greatest hits:

Who will mourn the passing of my heart?
Will its little droppings climb the pop chart?
Who'll take its ashes and, singing, fling
them from the top of the Brill Building

1 Elvis Presley in 1956

Love, death, despair and the pop song have been united as an inviolable configuration
of ideals, their chemistry maintaining a tireless, ever-renewing creativity. And this is a
process that begins, in the modern sense, with 'Heartbreak Hotel', as recorded by Elvis
Presley. Expressing terminal romantic despair in a way that became – paradoxically –
one of the most aspirationally glamorous statements in modern culture, the narrative
of 'Heartbreak Hotel' is shot through with a patterning of symmetrical symbolism, at
the centre of which was the image and reality of the heart – its capacity, sanctity and
vulnerability. It's a curious story, how 'Heartbreak Hotel' (just two minutes and eight
seconds long) became *the* song to inaugurate the modern Pop age, and in so doing
created a diversionary surge in the torrent of cultural evolution.

As the track that announced Elvis Presley as a major singing star, 'Heartbreak
Hotel' would have an alchemical role, turning the 'hillbilly cat' (as he had become
known on the gruelling country-music circuit) into the chart-topping recording artist.
And the lyric – dark, despairing, laden with fatalistic glamour – was no less strange
than the circumstances under which it came to be written. To borrow a weighty phrase
from Thomas Mann, there seems in 'Heartbreak Hotel' to be 'a mysterious harmony
between the Individual and the Universal Law' – some binding contract between the

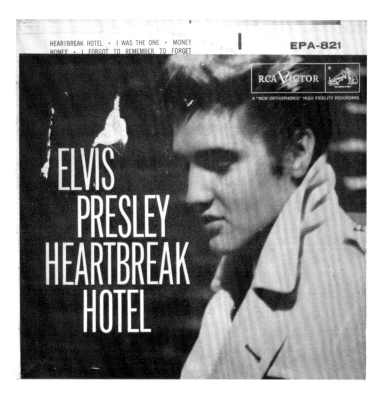

2 RCA record cover, 'Heartbreak Hotel', 1956

singer and the song, the terms of which describe not simply the life of Elvis Presley, but also the colossal resonance of his stardom.

It is the winter of 1955 in Gainesville, Florida, where one bright morning we discover in a cramped, unremarkable-looking office a small-time musician and song-writer called Tommy Durden, who is flicking through the morning edition of the *Miami Herald*. According to some accounts, Durden had only picked up the paper to study the day's form on the racing page, but what immediately caught his eye – right there on the front page – was the urgent, lurid picture of a corpse beneath a headline that screamed, 'DO YOU KNOW THIS MAN?' Like most of that morning's readers, Durden was immediately engrossed by such a morbidly compelling story. He read on. The suicide had removed all the identifying labels from his clothing, but had left a note that read simply: 'I walk a lonely street.' 'Empathetic' was the term that Durden was later to use to describe the way in which this tragic, enigmatic little episode, barely a footnote in the case histories of the Gainesville Police Department, seemed to reach out and touch his heart. 'A lonely street': the image was one of the suicide as somehow glamorous, poetic – fatally outside of regular society – the secular equivalent, of the Unknown Soldier; plus it had more than a tang of the blues about it.

We next cut to Durden, a little excited now, pulling up outside the house of his friend Mae Boren Axton. This is a woman with a portfolio of occupations – teacher, publicist, local radio personality, but at the top of the list, having got to know the young Elvis during his appearances in neighbouring Jacksonville, is the desire to write Presley's first big hit. Durden shows her the paper, tells her how that line – 'a lonely street' – just begs to be put in a song, and the two of them set to work.

At this point, we refer to Presley's hostile, despotic, self-tormented but keenest biographer, Albert Goldman – chronicler of these events – for the pivotal moment of the story. According to Goldman's account, it was Mae Axton – prompted perhaps by a more feminine response to the newspaper story – who imagined the family of the anonymous suicide suddenly seeing his photograph on the front page of the paper. How would they feel? How *could* they feel? The only word to describe it was 'heartbreak' – immediate and irredeemable sorrow.

We will never know whether the two phrases, 'lonely street' and 'heartbreak' came together in a single flash of intuition – 'Say, Mae! How about . . . ?' – or whether, in a fog of cigarette smoke, surrounded by screwed-up sheets of paper and emptied coffee cups, the two of them inched their way to the iconic line that would change not just the popular song, but Western society's experience of fame, glamour and youth. (And that of the East, too. Twenty-five years later, in November 1970, the Japanese author Yukio Mishima would also, in a gesture of terminal romanticism, commit suicide, in the wake of his failed take-over of the military headquarters in Tokyo. His chosen method – ritual self-disembowelment – was doubtless as corporeally unpleasant in reality as it had

3 Elvis Presley with Mae Boren Axton (right) and Charline Arthur

seemed heroically poised in concept. But when asked at a press conference, some months earlier, whom he would most like to be, Mishima had replied – without missing a beat – 'Elvis Presley!')

Back in Gainesville, Florida, in a simpler age, Axton and Durden were interrupted in their work by the arrival of what Goldman describes as 'a local singer and enter- tainer' from Jacksonville, one Glen Reeves. 'How about a little help here?', they ask. 'We're gonna write Elvis a hit.' Reeves asks the title of this big ol' hit they're gonna write. 'Heartbreak Hotel', they reply. And history records Reeves (in an act that would be matched by the manager who turned down The Beatles) reply: 'Weee-ll, that's about the silliest thing I've ever heard', before leaving to run some errands.

An hour later, the song was written and recorded by Durden on a little domestic tape recorder. Reeves had finally been coaxed into singing a version for the demo tape. Mae Axton then played it to Elvis at the National DJ convention in Nashville, where he immediately fell under the song's spell, exclaiming: 'Hot dog, Mae! Play it again!'

The other recordings made by Elvis during the period around his initial, spectac- ular ascent to stardom are almost without exception electrifying and compelling. As released on his debut album, 'Elvis Presley' (which did not include 'Heartbreak Hotel'), these would include 'Blue Moon', 'Tutti Frutti' and 'I'm Gonna Sit Right Down And Cry (Over You)'. But for all the allure, exuberance and erotic dynamism of these songs, it is 'Heartbreak Hotel' that seems somehow to define Elvis Presley as a modern myth, and present one of the great articulations of unbounded human emotion. Just what is it about 'Heartbreak Hotel' that would make it endure in a way that is different from these other recordings made by Elvis in this first golden period of his career?

Part of the song's impact must reside in the story contained in the song's central image – a hotel 'for broken-hearted lovers, to cry there in the gloom'. It is a lyric that combines modern urban glamour (lonely streets, crowded hotel bars) with figures who seem to step from the pages of some dark allegorical passion play: the weeping bell-hop, the desk clerk, 'dressed in black'. Such gothic imagery is also in keeping with the horror-story comic books of the mid-1950s, but it is the image of the 'Heartbreak Hotel' that does the bulk of the lyric's work – this established, the rest of the conceit takes shape with fluid ease. And at the centre of the song is the notion that to lose your loved one is to enter a kind of waiting room for death – a place where terminal loneliness equates to a kind of purgatory on Earth, in which regret and memory are the only surviving emotions. (We are reminded of Humphrey Bogart playing the ultimate romantic outsider as the lonely proprietor of Rick's Café in *Casablanca*.) Once broken, the heart no longer affirms life – the lyric suggests – it merely maintains the apparatus of living.

But there is also the sense that, as with other pop records that have become iconic – Kraftwerk's 'Autobahn' or Roxy Music's 'Virginia Plain', for instance – there is a sheer, radical newness to the sound and atmosphere of the recording that makes it not just mesmeric but also audaciously odd. Heard in the twenty-first century, a little over fifty years since it took the Number Two spot on the UK Top Twenty in May 1956, Elvis's recording of 'Heartbreak Hotel' seems now to reach us from a distance. Familiarity and enshrinement over five decades have rendered the track's originality all but inaudible; quickening, our hearing anticipates the contours of the song, yet simultaneously occludes the complexities – the oddness – of its performance. In the context of its times, however – a period when white American popular music comprised crooners, bee-bop, hillbilly and swing – Presley's 'Heartbreak Hotel' is without doubt a weird-sounding record. Faltering yet utterly assured, sparse but epic, insidious but cathartic, erotically charged but emotionally claustrophobic: small wonder that Goldman would write of the track, in 1981, 'a caricature of the blues, a sequence of melodramatic vocal gesticulations, it isn't so much a song as a psychodrama'. Elvis's declamation of the lyric's opening line – 'WELL SINCE MY BABY LEFT ME . . .' – rushes out to the listener from what seems like the silent emptiness of a pitch-black acoustic cavern. ('Before Elvis, there was nothing', John Lennon, in biblical mood, would later decree.) Underscored by a brace of deft, steely guitar chords, this lover's *de profundis*, in all its rallying urgency, simultaneously marks the beginning of an epoch – the Pop age of mass culture – and becomes en route paradoxically triumphalist.

4 Elvis Presley receiving gold disc for 'Heartbreak Hotel', 1956

The instrumentation, too, is full of surprises. What Sam Phillips, the founder of the Sun record label would call 'a morbid mess', is the result of a brilliantly tense arrangement (if something so raw can be called an 'arrangement') between Scotty Moore on guitar, Bill Black on bass, DJ Fontana on drums, and Floyd Cramer on piano. The whole emotional and erotic force of the track resides in its matching of eruptive outburst with teasing restraint; simultaneously prowling and swinging, minimalist and operatic, the recording takes us through a brooding bass line, ghostly bordello piano, slashes of switchblade guitar chords, to conclude in what sounds like a gothic reclamation of a vaudevillian flourish.

'Heartbreak Hotel' can be seen as a modern update on brooding, anti-heroic Romanticism – aspirational alienation for the age of mass media, as slick as the Cadillac tail fin designed by Harley Earl, and as poised as James Dean's performance in the preceding year's film portrait of juvenile delinquency, *Rebel Without A Cause*. In one sense, popular culture and pop music in the early and middle 1950s would both coincide with and contribute to the creation of the teenager as a new social group. Prior to the teenager, there had been children and adults, with a kind of No-Man's Land of sexual self-consciousness separating the two recognised states. The teenage phenomenon would of course be both defined and demonised as a grouping and a state of being that was perilously volatile – prone to violent mood swings, violence *per se*, and the full force of hormonally triggered melodrama. Small wonder, then, that within early teenage culture, the heart should be seized upon as a symbol of both the power and fragility of personal feelings – the slender line between swoon and doom.

Swirling outwards from the ancient lexicon of romance, 'heartbreak' is therefore a term precision-weighted to the needs of adolescence. In this sense, 'Heartbreak Hotel', as recorded and performed by Elvis when he himself was just twenty-one years old, became the perfect soundtrack (the 'psychodrama') for hormonally volatile teenagers – their mood constantly swinging between different forms of emotional excess, and traditionally finding considerable time to believe themselves doomed to an eternity of misery and isolation. Added to which, Elvis was extremely sexy: he made the perform-ance of despair appear erotic and heroic simultaneously, and with a sheer, animalistic violence that Frank Sinatra – the previous teen-idol – would never allow to disturb the sheen of his smooth urbanity.

There is therefore a secondary case for arguing that 'Heartbreak Hotel' was the song that invented the modern white teenager as this demographic would come to be defined by the commodity culture of mass production and mass media. There had been earlier contenders – not least the lachrymal Johnnie Ray, with his awkward good looks, and his bi-sexuality hidden more carefully than his hearing aid. And there had been film stars – polished to perfection by the Hollywood studios – and then Sinatra himself. But it was Elvis – matched point for point by his black counterpart, Little Richard – who would embody the whole idea of modern adolescence in a form that made all modern adolescents want either to be him, or be with him. And the key to his appeal was summarised – presented as its precious *extrait* – in the two minutes and eight seconds of 'Heartbreak Hotel'.

The extent to which Presley was summoning up some new, shocking, uncontrol-lable emotion from the hearts, first of southern country audiences, then of American teenagers everywhere, and subsequently the entire world, can be gauged from the collision between his earliest mature performances in 1956 and the responses of the authorities. Outwardly, at least, Elvis presented himself as a respectful, ultra-conservative, all-American boy; even as late as 1968, in the midst of American student radicalism, he would answer a journalist's demand to know what he felt about US involvement in Vietnam with the words, 'Ah'm sorry, suh – ah'm just an entertainer.' But beneath this flawless mask of duty lay an energy that refused to play nicely. The heart of white southern American morality in the mid-1950s was both paced and defied by the heart of white rock and roll that pounded through 'Heartbreak Hotel'.

Jay B Leviton's photographs, collected in the book *Elvis Close-Up*, powerfully record the collision. At Presley's shows in Jacksonville – which is where all that screaming really started in earnest – seats were also occupied by Juvenile Court Judge

5 Elvis fans at Florida State Theatre, Jacksonville, 1956, photo by Jay B Leviton

Marion W Gooding and an unsmiling squad of what were termed 'morally upright committee members', there to see for themselves this potential threat to Jacksonville's youth. All went well as the warm-up acts played their sets: nice country songs, some polite applause. But when Elvis walked out, the screams began, and he'd hardly got as far as singing 'WELL SINCE . . .' when all hell let loose.

The judge summoned Elvis for a cautionary warning – Presley's act must be toned down. But Leviton's compelling photographs of this encounter reveal that the judge

6 Elvis Presley performing at Florida State Theatre, Jacksonville, 1956, photo by Jay B Leviton

7 Judge Marion W Gooding (right) in audience at Florida State Theatre, Jacksonville, 1956, photo by Jay B Leviton

is playing a losing hand. In what resembles a father-and-son pep-talk, Elvis is photographed listening attentively to Judge Gooding; it was impossible, however, by this stage in his burgeoning career, that these words of guidance from a senior member of the community could stem the near-elemental power of Presley's act – or his popularity. By these heady months in 1956, Elvis had become the first truly modern mass-media phenomenon. His aura is palpable in these pictures – like a halo of brightness around his perfectly quiffed hair. What Leviton got were photographs of youth versus old age, acceleration versus stasis, old versus new. In his book *Time Travel* (1996), Jon Savage defined the concept of acceleration in relation to pop: 'Speed is vital because it is one of the few areas where teens are more powerful than adults, and time has frequently been used to express a rebellious attitude . . . I'm faster than you: I'm five years ahead of my time. This is a generational war, expressed in time and perception.' In these terms, Elvis is simply *faster* than the judge.

But this is not to belittle Gooding with the superiority of hindsight; for the judge – well-intentioned, doubtless – was aware of a growing national fear not just of juvenile delinquency, but of unchaperoned and unmarshalled youth *per se*. It was a situation that the exploitation films and novellas of the 1950s would seize upon with ravenous eagerness: pornography disguised as edifying sociology. But it was also a trend that Truman Capote, with terrifying lucidity, would sum up in his forensically detailed reconstruction of a quadruple murder on a remote Kansas farm, *In Cold Blood*, a scenario that

8 Elvis Presley and Judge Marion W Gooding, 1956, photo by Jay B Leviton

Capote would summarise – to paraphrase – as describing two Americas: one, the safe, respectable civilised America, and the other, running parallel but perilously over-lapping, where they'll slit your throat as soon as look at you. 'They' were the murdering hoodlums whom folk like the judge feared were not merely represented but actively encouraged by a phenomenon like Elvis.

Perhaps this was not so far from the truth. Presley's performance of 'Heartbreak Hotel' represented the unleashing of the normally repressed, ultra-violent, sexually voracious capacity of the human psyche. In his recording of 'Heartbreak Hotel', there is a quality to Elvis's singing that seems entirely intuitive – straight from the heart. For Goldman, however, the very strangeness of 'Heartbreak Hotel' is due to the fact that Elvis was simply copying – as best he could – the way in which Reeves had sung the song on the demo tape, quite intentionally, *in the style of Elvis Presley*. Thus Elvis was copying an Elvis impersonator.

If this were in fact the case, then 'Heartbreak Hotel', as it subsequently soared up the US and UK pop charts, would announce the Pop age in terms that were purely and classically postmodern: with appropriation, punning and deconstructed authorship. All of which would be highly inviting to accept, in terms of cultural historical neatness, save for the fact that it was Reeves who was copying Presley's already distinctive vocal style in the first place. What does emerge from the Goldman theory, however, is the fundamental dislike that Goldman seems to have for his subject – rather as though, if he can discredit Elvis Presley as an icon and as an entertainer (starting with the suggestion that he couldn't even sing), he will have somehow purged American culture of a disease in its own heart. Goldman and Judge Gooding become virtually interchangeable in their intention, if not their motivation: Elvis must be brought down before he gets above himself.

In one sense, Lennon's announcement that 'before Elvis there was nothing' was entirely accurate. For in terms of celebrity, as well as in terms of his musical career, Presley was like one of those unfortunate animals sent into orbit by the US space programme towards the end of 1950s. Nobody had actually been where Elvis was going before; his degree of fame, and its acceleration were entirely unlike the controlled fame enjoyed or suffered by film stars. In Hollywood, there was a method and there were rules – F Scott Fitzgerald had observed them first hand, and written their constitution (official and not so official) in his final, unfinished, great novel, *The Last Tycoon*.

There is a way in which Presley's career – like that of Oscar Wilde, or Billie Holliday, or Scott Fitzgerald himself – seems every bit as scripted as the doomed romanticism of 'Heartbreak Hotel'. The young Presley pursued an almost vertical ascent, to reach a premature apex. The rest, the downfall, is pure Aeschylus. For just as Elvis had begun by embodying and performing the modernity of the Pop age, so he would fall victim to the compensatory pleasures – in his case drugs and food – that would come to define the shadow of the new consumer culture. The phrase 'pop will eat itself' might well have been coined for Presley, and herein lies his own tragedy, and his own heartbreak.

The American music writer Greil Marcus, in his bravura meditation on Presley, *Dead Elvis: A Chronicle of Cultural Obsession* (1991), investigates at length what might be termed the 'afterlife' of Elvis as a cultural phenomenon. In examining Presley's posthumous existence as a kind of pan-cultural phantom, haunting the modern

9 Heart-shaped pool at Heartbreak Hotel, Gracelands, Memphis

consciousness, Marcus pronounces that he has become a 'necessity' – 'the necessity existing in every culture that leads it to produce a perfect, all inclusive metaphor for itself'. In this much, 'Heartbreak Hotel' might well be regarded (and Jung makes the point that the creative self is imbued with the gift of prophecy) as an act of premonition. The song does inaugurate the Pop age of mass consumerism, does announce postmodernism, and is in fact both requiem and magnificat for a society *in extremis* – the end of the world starts here. (John Cale would later record and perform the song as a virtual dirge, shot through with spookatronic synthesiser effects created by Brian Eno, and backed by three female vocalists who sound like a Tamla Motown girl group by way of Edgar Allen Poe.)

As a ghost within the culture, Elvis undergoes a kind of posthumous transubstantiation: his death from heart failure and cardiac arrest – consequent on his extended addiction to drugs and food – transforms him into a symbol of both eternal youth and eternal decay. For the punks of 1977, he was presented as Pop's first zombie; for the post-postmodernists of 2007, he is an item among millions on a database of endlessly recycling cultural styles.

As one of the starting whistles for both the Presley myth and the 'total Pop world' that Andy Warhol observed had surrounded American teenagers as early as 1962, 'Heartbreak Hotel' now takes its place as one of those creations in which we see ourselves reflected. What Tommy Durden read on the front page of the *Miami Herald* all those years ago, became translated by the song that he co-wrote, and that Elvis Presley recorded, as a recognition of mortality. If a man could walk a lonely street, leaving no name and just a line of scrawled writing, then somewhere in his tragic tale was a quality that – made eloquent by song – resounds in the breadth of our common contract with life.

INTERVIEW
Chris Rodley

Chris Rodley

Film director, in conversation with Melissa Larner

MELISSA LARNER: When did you first realise there might be something wrong with your heart?

CHRIS RODLEY: I think I'd always expected heart problems, because it's in my DNA. My father died of a heart attack at fifty-two, and my mother had a heart condition from the age of fourteen, when she had rheumatic fever. I was very young when my father died. He was 6 foot 4, 9 stone 10, a London-to-Brighton walker, and to all intents and purposes very healthy. He'd been to the doctor three days before he'd died, with pains in his chest, and was told it was indigestion and to go home. So when he got up on a Saturday morning, aged fifty-two, and died of a heart attack, it was a big surprise. I don't think even my mother knew there was anything wrong with him – it was a great shock to her. She just glommed on to this phrase that was on the death certificate: 'Coronary thrombosis due to atheroma'. I don't think she knew what it meant, but it became a kind of mantra in our house. And she was often ill with her heart. So I think I always thought a long life wasn't necessarily an option for me.

ML: And did you get any warning before your heart attack?

CR: Yes, I did have some chest pains. I wondered if I'd had a heart attack and just not realised it at the time, so I went to the hospital and they did blood tests and they said, 'No, you haven't had a heart attack.' I think they can tell through the enzymes in the blood. Stupidly, I didn't ask, 'Do I have angina?' And stupidly on their part, they didn't investigate that. They just gave me a clean bill of health.

ML: And then what happened?

Pradeep Narayan, Registrar to Professor Gianni Angelini at Bristol Royal Infirmary, 2005, digital photo by Mike St Maur Sheil

CR: Well, two years later came the heart attack. It was a Saturday morning, just like with my father, and I was roughly the same age – nearly fifty. And I'm a carbon copy of him anyway, so it was like a sense of destiny. It was sort of not a surprise. I just thought things were being played out as I always knew they would. It sounds odd, but that sense of destiny took some of the fear out of it. It just felt like I'd already glimpsed the last page of the book and now I'd actually reached the end.

ML: Are you saying you didn't feel frightened?

CR: I'm not sure fear was actually the first thing I felt. It was a bit like the last line of Daphne du Maurier's *Don't Look Now*: 'Oh bloody hell' – when the man's killed by the murderer in Venice. He felt stupid because he'd seen it coming, but hadn't avoided it. And I was also thinking about Glenn Ford, clutching his arm in *Superman*, dying in a cornfield on a farm – all those sorts of images flash in front of you. But it was more like a sense of embarrassment, because I knew it was going to be a problem for everybody. My girl-friend and I were leaving for Paris that morning. So it was a feeling of embarrassment, of inconvenience, and 'Oh here we go – I'd rather it wasn't happening today.' And I was also thinking, 'Is this really happening? Is that really what that pain is?'

ML: You were in a lot of pain?

CR: Yes, a lot of pain. It was all the classic symptoms – pain down my arms and across the chest. And then I thought – and maybe it was a hangover from my father – that it could be just indigestion. You don't want to alarm people unnecessarily; you want to be sure that that's what it is, so you're rather tentative about it. I mean, some people think they're having heart attacks when they *have* just got indigestion; some people are terrible hypochondriacs. I wanted to be absolutely sure that this was it. There's a fantastic billboard at the moment, with a guy with a belt tightening across his chest, and the message is, 'If you have chest pains, call 999.' But you sort of wait because you're about to go to Paris and you don't want to spoil it. It's daft. You're thinking, 'What if we don't go, and it was just indigestion?' And I suppose you're also hoping that's not what it is. But eventually I said to my girlfriend, 'I *think* I'm having a heart attack', and she called the ambulance and made me lie down. And they came very quickly.

ML: How did they treat you?

CR: I remember very clearly, the first thing that came out of the paramedic's mouth was, 'Have you been taking Viagra?' That was the first thing he asked. [Laughs] I assume that must be a new thing they have to ask. And then it was just describing what it felt like. And after that, they take you out and wire you up to some machines in the ambu-lance.

ML: And what were your feelings then? Did the fear kick in?

CR: Not really. They came very quickly, and you're grateful, because people are taking care of you. I absolutely believed that they knew what they were doing, and that it was being dealt with and therefore it might be alright. And the pain wasn't getting any worse. And they're quite practised at being chirpy and not alarmist. They're matter of fact and helpful, like a plumber – 'It's just a blockage; we'll fix it.' I think the worst thing about being in the ambulance was that for some reason, they didn't let my girlfriend in too. There was a blackened window. She couldn't see in, but I could see out. And I could see the look on her face as she was pacing about outside. I think it must have been much more worrying, believe it or not, to be outside the ambulance. That's something that happens right the way through: the look on other people's faces. You want to tell them it's OK – whatever they're imagining, it's worse than you're experiencing. And you can't really do anything about it; you can't convince them that it's actually not as bad as they think it is.

ML: So you didn't think you were going to die?

CR: I just didn't know what was going to happen. But no, I didn't really. I thought, 'It would be worse than this if that were the case, and anyway, I'm in good hands.' So I didn't think about dying. Maybe I just *couldn't* think about it. And it's OK as long as there's activity going on and things to distract you – people doing things and asking you questions. Maybe they do that on purpose – they ask you stuff all the time. If you were on your own and it was happening to you, it would be really horrible, but as long as there are people being purposeful and asking you things and plugging things in, and shooting stuff into your arm, it gives you something else to focus on rather than some bleak end to it all.

ML: And next they took you into intensive care?

CR: Yes, at the London Hospital and they did some kind of test on me to see if I'd actually had a heart attack. That was weird. They wire you up to the machine, and a print-out comes out and you look at their faces, and they look at each other and you just know. There's no hiding the looks on their faces. And they're nodding and consulting, and you're deciphering what's going on from the look on their faces. And then they told me: 'You're having a heart attack.' And they asked me to make a decision very quickly, which was a bit scary, because I'm a terrible procrastinator and can't make decisions about anything, except in my professional life. They said, 'We can give you a clot buster, but some people die when we give it to them. What do you want to do?'

ML: The risk there is that you can have a stroke. But was there really much of a choice? Or did you feel you were being strongly advised to do it?

CR: There probably wasn't any choice. I guess it's just legal stuff. But you're not thinking like that at the time. It felt as if I had to make a BIG decision, even if they were just saying it to cover their backs and death was unlikely.

ML: So they gave it to you?

CR: Yes. And then the only disappointment was that they shot me full of diamorphine and said, 'You're going to feel lovely', and I didn't feel lovely. I didn't feel any better mentally, but the pain stopped.

ML: Can you remember how long you were in intensive care?

CR: I don't think it was that long, actually. An hour? And then I was wheeled out and taken onto the general ward. I think I was the youngest on the ward, although someone younger came in later. Everyone was much older and seemed much sicker than me, or maybe it just felt that way to me. There wasn't a lot of pain at that point. They were just monitoring me. There were little moments of pain, but I felt like a bit of a sham. I felt it wasn't as bad as they thought. At that point, I didn't have any doubt that I was going to walk out of there one way or another, although there was someone in the ward who, even over the short period that I was in there, came in OK and went right down. I don't know what happened to her.

ML: Surviving the night after a heart attack is the next hurdle. Were you scared of that?

CR: I don't think I knew that the first night is a crucial one. I'm glad I didn't!

ML: And you were fine the next day?

CR: Yes. And then the cardiologist came to see me. And he decided that they needed to do an angiogram.

ML: What sort of relationship did you have with him?

CR: It was odd, because I'd never met any kind of consultant before, and then it struck me that they're like stars. He came in like a VIP. He was a bit flash in the way he dressed. Actually, I was going to have a word with him about his clothes, because he was dressed rather funereally. [Laughs] He wore a dark suit, a size too large – like a member of the Addams family. He was quite young, but his suit just spoke of a dark joke to me. I gathered that he was a bit of a whiz kid, and he was a propagandist for the radial method of doing angiograms and angioplasties – through the wrist rather than through the groin. The artery in the groin is like a six-lane highway and much easier to access, but it means you have to lie down for a few hours after the procedure, whereas the great thing about the radial version is that they just put a clamp on your wrist and you can get up and walk away. It's trickier – I guess the passage is finer

– but it's a shorter route. He was saying that we should be moving towards radials. So, if you're like me, you sort of hope that because you're going to be a statistic on a medical paper that's trying to prove the superiority of a certain kind of treatment, then you're in good hands – they really don't want it to go wrong. You're part of another agenda. But we were alarmed when we were told about the procedure because they said that there was some risk attached. The problem is that the consultants don't really feel that they owe you a full explanation. You're supposed to be incredibly grateful – which you are. They're going to save your life, and you're not even going to have to put down your credit card. I became a major propagandist for the National Health Service afterwards. But they don't seem to expect you to ask too many questions. And he wasn't a great communicator. I remember he didn't address any comments to my girl-friend when he came to the bedside while she was there. They do have that attitude where they don't necessarily explain anything. It's like having a plumber round – they don't feel obliged to tell you all the details, because a) it's none of your business, and b) it might be a bit technical – you wouldn't understand it. But basic psychology is that you do explain things, because if they only tell you bits of it, it's opaque and therefore a bit frightening.

ML: He didn't explain the procedure?

CR: He sort of explained what was going to happen, but it just seemed like science-fiction and not possible, and it sounded much more invasive that it actually was when only partially explained. We were told there was a risk that I might arrest while on the table. And it seemed like quite a high statistic. Or it's high to someone who does the lottery every week and thinks he's going to win at thirty-seven million to one! [Laughs] In fact, a guy did arrest on the table, while I was waiting to have mine done. He survived it OK, though. They shot him through with adrenalin. He was an otherwise extremely fit young fireman. And everyone in the ward sort of high-fived, because we thought, 'That's us off the hook. He's the statistic!' It's like if you don't want to get on a plane with a suicide bomber on it, take a bomb on the plane yourself, because the chance of there being two bombs on the plane is almost zero. Anyway, he was absolutely fine. But I felt sorry for him, because he had no job to go back to afterwards – his career was finished. Actually, most people on that ward didn't want to work, couldn't work, or were never going to work again. So I think they were a bit puzzled when I kept saying, 'When can I get out of here and get back to work?' I was half-way through the production of a film, and they said, 'Well, you're insane.' I was asking, 'When can I get on a plane again?', because we had to go to Greece to film. They said I had to wait six weeks and I did it in seven. I just wanted to get back to work.

ML: Can you describe the angiogram?

CR: Well it was an angiogram that became an angioplasty, because they went in and said, 'Whoops! We'd better do it now.' They knew it was a pretty serious blockage, but it turned out to be 99 percent blocked.

ML: You were very lucky you were able to have it fixed with an angioplasty, with such bad occlusion, rather than with a bypass.

CR: Yes, and again, I began to feel a bit of a sham. When I learnt that they do five or six of these things a day, five days a week, I thought, 'It's like having a tooth extracted. It's totally routine.' They've been doing it for about ten years, and obviously they've got better and better at it. I think St Bart's, where they moved me from the London Hospital, was particularly good.

ML: You're kept awake during this procedure, aren't you?

CR: Yes, I liked that. I'd never been put under before, and the idea is horrible to me. You have to be awake, because they need to be able to ask, 'Can you feel this? Can you feel that?', or 'How are you feeling now?' Parts of it are painful, and you're supposed to say, 'I'm feeling pain now.' It's like a sort of prayer – like Roman Catholic responses. And then they say, 'Great, that's exactly what you should be feeling because we're inflating the little condom thing to open the artery up and you will feel some pain. Count down from ten and it'll be gone.' And I counted down, and it went. So, it was like you're the car and if a car could speak, it would say, 'My spark plugs need fixing', and they say, 'How's that?', and you say 'Oh, that's better.' And I quite liked being awake because you kind of get lost in the mechanics of what the cardiologist is doing.

ML: You're watching it on a screen, aren't you?

CR: There is a screen there, but I couldn't really look at it, and they try to conceal their operations as much as possible. I think the weirdest thing is as the catheter goes in – a tube goes up first, and then there's the instrument that goes inside that, and you can kind of hear it scraping. Or you can feel it, and you imagine that you hear it.

ML: That sounds horrible.

CR: But when you go through something like this, you just become like a pincushion, and you get used to it. A catheter goes in here, and then another thing goes in there, and you don't really mind because you know it's all good for you. I got to the point where if someone wasn't sticking tubes in me, and if there wasn't blood and swabs and bruising, then I thought something was wrong. I've always been squeamish about injections, but you just become, 'Yeah, whatever. Do your worst.' You have to find another level of tolerance, which you do very quickly. You can't be squeamish.

It was also in some ways of benefit to me that I was in a bed next to a young girl who'd been crushed by a horse. Her horse had fallen on her, and she was virtually a broken china doll. She was in pieces and in a lot of pain. She was on incredible amounts of medication. It was horrendous. She was in a really bad way. So I had nothing to complain about.

ML: And in order to open up these very occluded arteries, did they have to put in those metal stents?

CR: Yes. The cardiologist calls out for these stents of varying lengths. The left artery is particularly crucial, and with that one, I only had one stent, so I was on and off the table quite quickly – about an hour. And then he showed me the pictures: what it was like before and what it was like now. And what it was like before, was just like one of those balloon animals you twist for kids – the tube went to nothing! And then, there it was afterwards, all flowing normally. So that wasn't a lot of trouble. I got up and walked away.

ML: And you went home after that?

CR: I went home, but within two or three days, I started getting pain again. A lot of chest pain. They'd said when they did the left artery, 'We'll do that one; that's the serious one, and then we'll do the right one some time in the future.' It was very vague; there was no schedule. But then they decided I needed the second one as soon as possible.

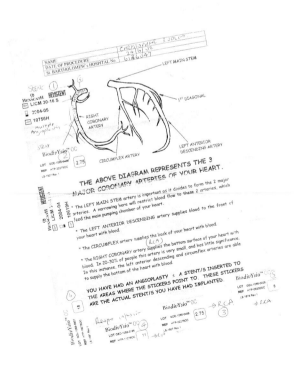

Diagram showing coronary stents implanted during angioplasty

when they did that one, the shock was that they put seven stents in, and I was on the table for over three hours. That was more traumatic, because when you're on the table for three hours, you run out of gags and you run out of patience. And then you hear the cardiologist say, 'Get me another 18 cm' , and you go, 'Oh god, another one? That sounds really long!' I just couldn't believe that he kept asking for another one, and then another one, and yet another one. I said, '*How many are going in*?' And I did start to panic slightly because during the course of three hours, all those things about arresting on the table hit you. It's a long time to be lying on your back just looking at the damn ceiling. You do start to imagine all kinds of things.

ML: Was that the worst part of it all?

CR: I think the worst part was that I felt reduced to an organism – to elements, to components. You're having an identity crisis anyway, because you know your life's supposed to change. You get read the riot act about your smoking, your eating and your drinking. Some people really seem to delight in that. The nutritionists really delight in it – they were particularly horrible. They were gloating that I was going to be just like them – sucking on pebbles for the rest of my life! [Laughs] I wasn't going to be able to have all that lovely pork fat anymore. So they're asking you to do a lot to change your lifestyle, and at the same time, you just get reduced to a kind of organism. I didn't even want my friends to come and see me because I didn't want that to be my identity. I didn't want them to see me in that context. My heart's been broken many times over, but it wasn't the heart attack that broke it. The heart's just a pump. That's not what matters. Wherever you reside as a person, an individual, it's not there. Unfortunately, though, when it ceases to work, then you cease to be, so in hospitals you're just reduced to the things that need to be fixed, even though that's not where you reside. So you just feel like a pertri dish when you're in there. They're not dealing with *you* – you're a thing with a blocked left artery and a screwed-up right artery. And you just become that organ. That's a really odd experience. It's not a criticism. All the hospital staff were nice, and I could have a good laugh with the nurses. What else are they supposed to do? If you want meaningful chats, go to your therapist. They can't do all that. The cardiologist comes in and you're the job of the day, and he doesn't want to know about you. And why should he? But you sort of *want* them to know about you. It's terrible – you want to say, 'Hello! Here I am. This is me. Can I not be just this heart?'

ML: But you must be grateful as well. They saved your life.

CR: Oh yes. I wouldn't go private if you paid me. I think these people are amazing and they do a fantastic job. And it's weird, because when I had to go back for check-ups, I really wanted to see my cardiologist again. There's been this intimate moment between you. He's been inside you. I went back and saw him, and there he was, that guy who saved my life, sitting in the same funereal suit, and I found I was really pleased to see him. It sounds weird, but it was very reassuring, and I didn't really want to leave. I wanted to sit and chat. I thought maybe we could be pals! [Laughs] But of course, he's got a million people to see and he really just wanted to say to me, 'Don't ever smoke again, mate.' And the next time I went back, it wasn't him. It was his second-in-command and I felt quite crestfallen, even though she was much better at explaining things. She told you exactly what you wanted to know, and wasn't so grand. So actually, I understood a lot more from her. But I wanted to see *him*, because he was the guy who saved my life. You can't help but have that strange, extremely temporary but intense relationship with someone who's been through it with you. And I haven't deleted his number from my mobile phone. *Why?* Why have I got it on there? I've never called him and I never will, and I hope I never have to see him again. But I haven't taken his name off there because he was the guy who saved my life. It's odd, isn't it? To me, it's everything; to him, it's one of six in a day.

ML: When you came out of the hospital did you feel anxious?

CR: Yes, but I didn't want that to be my identity. We went straight into filming and the lovely assistant-camerawoman bought a little fisherman's chair, one of those fold-out things, and was always insisting that I sat down between shots. I thought, 'I really don't want it to be like this.' I'd always lugged equipment and it was all hands to the pump, but they wouldn't let me do anything. I thought, 'Well maybe I shouldn't', but I also thought, 'I hope this isn't how it's going to be from here on in' – which in fact it hasn't been.

ML: And four years later, you're fit and well. Do you think about your heart attack much?

CR: Every day. They said, 'You could live another thirty years.' But I still think about it every day, and many times a day. It's something that hangs over your head.

ML: Do you feel the experience has enhanced your life in any way?

CR: Well, your life changes – I don't know about enhancing it. It's an add-on. I think there are good things about it. You can reassure people who are worried about being in a similar situation. One of the nicest things in hospital was going for the second procedure, and there was an old guy waiting for his first angioplasty, and he was so frightened. He said, 'I've fought on the battlefields in World War II and I've never been this frightened.' And it was really great to be able to say, 'It's a piece of piss. I've had one already. You just lie on the table; it's easy.'

So it changed my life, and it's not a matter of for better or for worse: your life is just something else afterwards. You're less scared. You feel a bit toughed up. I can't think why people do dangerous things like climbing mountains and taking those sorts of silly risks, but my equivalent is having survived a heart attack. You get to the top of the mountain, and you live to tell the tale. And you can say, 'I've been there. It's alright. I survived it.'

Sick at Heart

Appetite Versus Technology in the Twenty-first Century

Jon Turney

Eat Your Heart Out

The new millennium has brought an awareness in affluent societies that our collective appetites threaten our future. Globally, signs of climate change add up to a compelling case that we have limited time – perhaps only a decade or two – to change our ways before Really Bad Things start happening worldwide. But the news for individual health may distract us from the prospects for the planet. Our compulsion to overeat, to gorge on lovely, nasty fats and sugars, is creating generations who have a lot of extra body mass to haul around. As well as the physical load, the epidemic of obesity promises to add to the burden of disease. And along with diabetes and strokes, the millions of beneficiaries, or victims, of our calorific cornucopia face a future with a scarily high incidence of heart disease.

As with global warming, the solutions to the problem of eating ourselves to death fall, in principle at least, into two classes. Maybe we could change our behaviour to turn the numbers round. Failing that, there might be a technical fix that would compensate for our weakness of will by saving us from the consequences. It is only sensible to explore both options. But since the stakes are high in both the planetary and personal realms, backing the technical fixes might be the more attractive bet.

Admittedly, the analogy is imperfect. Heating the planet affects everyone, the virtuous along with the gas guzzlers, and individual action to reduce one's carbon footprint is easily overwhelmed by others' less prudent habits. Avoid obesity for yourself, though, and you get all the benefit. It is true, too, that the recent news on heart disease is, well, heartening. The death rate from cardiovascular disease has been reduced by

1 Our compulsion to overeat is creating an epidemic of obesity, causing diabetes, strokes and heart disease. In the UK, coronary heart disease remains the largest single cause of death in men.

around half in the last forty years in most developed countries, Britain included. OK, we'll die of cancer instead, but this is still an astonishing fact. The main reasons seem to be a whole series of improvements in managing cardiovascular disease, the success of anti-smoking propaganda and, latterly, the introduction of cholesterol-lowering drugs.

On the other hand, the reduction followed a period when the incidence of heart attack was staggeringly high. The death rate from coronary heart disease among the under-sixty-fives in England in 1970 was eighty people in every thousand. And forty people in every thousand still fail to collect their pension because their heart gives out before then.

So the success is tempered by a sense of work still to do. In the UK, coronary heart disease remains the largest single cause of death in men, at 21 percent in 2004, and in women, at 15 percent. (Cancers of all kinds account for more, but constitute a collection of different diseases.) If one reads the British Heart Foundation's exhaustive annual compilation of coronary heart disease statistics, it is hard to be optimistic about this changing.

The reason is that while one behavioural change, giving up the smokes, has helped, another set of changes in habit seems to be undermining the benefit. Although advice about diet and exercise is deployed with a regularity well beyond tedium throughout education, it is largely unheeded. And recent surveys suggest a rise in obesity and a disinclination to exercise among the young that worries public health strategists as well as health-service budget holders.

Another way of putting it is to suggest that we in Britain will become, on average, more American. A survey published last year (2006) in the *Journal of the American Medical Association* indicated that middle-aged (fifty-five to sixty-four) white people in the USA have twice as much diabetes, higher blood pressure and 50 percent more heart disease than their British contemporaries. The researchers found the difference surprising, but noted that the rise in obesity began earlier in the US. The incidence of obesity there increased from 16 percent to 31 percent between 1980 and 2003, while the UK figures only went from 7 percent to 23 percent. As this suggests, though, the Brits are catching up.

Enough numbers. The problem has been widely recognised, and calls for action are equally widespread. Will they succeed? It seems doubtful. Stopping smoking meant relinquishing a pleasure that takes effort to cultivate, and which, relatively speaking, is historically novel. Eating junk and loafing are older pleasures. Or rather, trashy modern processed food is tailored to ancient evolved appetites that are harder to control. And new pastimes also provide rewards while sitting in front of a TV or computer screen that might once only have been obtained by leaving the house.

The health educators will go on trying to mend our ways, but the companies promoting these things, especially the junk, are creative, resourceful and relentless. As I write, my newspaper catalogues new 'novelty foods', including Bernard Matthews Turkey Princess Dreams, Sainsbury's Triple Chocolate Crisp Cereal, KFC Popcorn Chicken and Kellogg's Coco Pops Straws. Do the healthy eating advocates have temptations like these? No chance.

So let's assume that heart disease is going to get worse again; that the most vital muscle and its blood supply will need help more often in an increasingly obese, hypertensive, diabetes-ridden population. If changing behaviour falls short of achieving heart health, could there be that technical fix?

How Far Can We Go?

The heart is a more complex organ than many appreciate, with its chambers, valves and co-ordinated electrical control of all the muscle cells, which work together in rhythmic contraction. But its task is a relatively simple one. It is, as Harvey appreciated, a pump. So perhaps it makes sense to think about three strategies for keeping the pump going – easing the flow, repair, or replacement.

There are improvements in prospect for all three. Some are fairly straightforward extensions of existing treatments. Others, inevitably, are more speculative; the heart and circulation often figure in the most speculative future visions of any in medicine – in which nanobots (microscopic robots) scour our arteries, clean and patch up damaged cells. But anyone who thinks these are going to be working away inside us any time soon, or that we could devise control systems to ensure that they actually did what was needed, is operating on the margin where technological optimism tips into fantasy.

While we wait for the nanobot cavalry, more and better drugs offer some hope for both treatment and prevention. The old familiar aspirin is now firmly established as a benefit to people at risk of heart attack – especially when the risk was established by having one. So are various clot-dissolving drugs and enzyme inhibitors. But everyone's favourite class of drugs at the moment is the class known as the statins, which lower blood cholesterol. These drugs, first discovered in the 1970s, cut the level of low-density

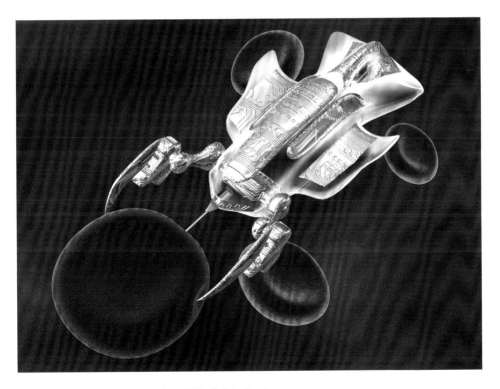

2 Oliver Burston, imaginary nanobot, 2005, digital artwork

lipoprotein cholesterol (LDL-cholesterol, the most harmful variety) by blocking a synthetic pathway in the liver. Taking them cuts blood levels, and heart risks, for people with elevated cholesterol. But it now appears that it also reduces risk for people with 'normal' cholesterol levels. This begins to look like a pharmaceutical executive's dream – a drug that everyone should take, all their lives, even if they are not ill.

There are hints that the beneficial effects of statins are due to anti-inflammatory properties rather than reducing cholesterol – and that fits with theories that arterial blockage is at least partly caused by inflammatory reactions. No matter, the statistics show that something good is going on. There will be diminished benefits at some point, if only because a person with no cholesterol at all would have very peculiar cell membranes. And statins may not be much better than aspirin for those whose cholesterol level is normal, but the drugs are certainly indicated for people with high levels. They already constitute the most widely prescribed and most profitable class of drugs. Whether they are potent enough to stem a rising tide of cholesterol in the veins of a new generation of fat-addicted couch potatoes remains to be seen.

Fixing the Machine

Beyond drugs, we have to get at the heart more directly. Here, there is another distinction worth making. The three strategies each have a range of technological possibilities. Some are biological, some rooted in more traditional kinds of engineering. As we shall see, they are not mutually exclusive, but describing them separately makes it easier to follow what is going on.

Easing the flow is now a very common surgical procedure. If arteries, especially the coronary arteries that furnish the heart muscle's own blood supply, are getting clogged beyond the stage where drugs can help much, then surgeons go in and widen them by pushing the walls apart. This delicate procedure is made possible by a whole range of tiny wires, catheters, balloons and stents. The basic procedure is to insert a small tube into the blood system, usually by making a hole in the femoral artery. This is then channelled along until it is in the right place for the real work. Originally, this procedure was used in order to put in place a balloon that was inflated to widen the artery at the site of blockage. Today, the end of the catheter tube can carry an ever-increasing set of miniature devices – tiny cutters, laser guides, cameras, echocardiographs. The balloons are often replaced with stents – small tubes that sit inside the artery, keeping it open for the flow of blood. Research now, which does involve nanotechnological tricks, is finding new ways of impregnating stents with drugs, which are gradually released to help stop the new tube eventually constricting in the same way as the artery did in the first place.

Beyond patching up old pipe work in the arteries lies the challenge of repairing heart muscle. Here, the rosiest prospect is biological. Congestive heart failure can arise

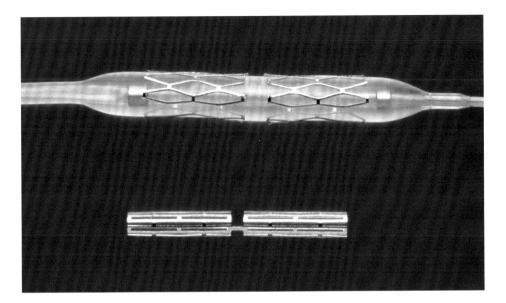

3 Articulated coronary stent

from sustained high blood pressure, poor blood supply because of silted up coronary arteries, or the thing that really gets your attention – a heart attack. In each case, heart muscle cells, cardiomyocytes, die – leaving the heart too weak to do its job properly. There may be limited growth of new heart-muscle cells after injury, but hopes for real progress currently rest on stem cells, the much-discussed hope for curing a range of conditions whose symptoms are due to cell depletion.

The idea is simple, the practice often exquisitely complex. Cells exist that can give rise to all the other different kinds of cells in the body – after all, we all began from a single cell. They are found in embryonic tissue, although some, nearly as developmentally potent, persist in adults. If they could be coaxed into replacing heart cells, whether myocytes or the endothelial cells or smooth muscle cells that make for healthy blood vessels, they could offer a bottom-up route to repair that is far more accessible than the top-down method of a heart transplant.

Most of this work is still confined to mice and rats, where both rodent and human stem cells, some from bone marrow and some from embryonic cell lines, have shown good signs of behaving as researchers hope. That is, if a heart is damaged, stem cells injected into the right place will migrate to the right regions of heart muscle and replenish the population of muscle cells with new ones, which join in the orchestrated contraction that makes up a heartbeat.

Various trials on patients are now getting under way, most using their own stem cells from bone marrow, but there is still much to learn about how the movement and other responses of stem cells from all sources are controlled. In principle, though, cell therapy has a great future. Evolutionary studies also encourage the belief that stem-cell therapies can be made to work. Amphibians, fish, molluscs and arthropods can regenerate cardiac muscle in their (less complex) hearts, but warm-blooded animals have lost this ability. A detailed understanding is now emerging of the gene-regulatory networks that appear to have remained in place, with elaboration and modification, since the first proto-heart appeared some 500 million years ago. It may mesh with stem-cell work as researchers try and guide developing cells along particular pathways.

Also important in the background to this work will be the overall understanding of how the heart's movements are co-ordinated and controlled. It is often said that bringing together all the data now accumulating about genetic sequences, protein synthesis and metabolic profiles will depend on the advent of a new, integrative 'systems biology'. Examples of how this might work have tended to feature single-celled organisms like yeast. But when it comes to human organs, the nearest we have yet come to systems biology may be in studies of the heart. The work of Denis Noble and colleagues in Oxford University, for example, relates information about the heart at many different levels through computer modelling of heart-muscle cells. This began decades ago with attempts to model the behaviour of one cell, and how it responds to electrical stimulation. Now the models have been scaled up, with contemporary computing power, to give an enormously detailed, dynamic representation of how the heart behaves.

To take just one example, it is possible to bring together knowledge of variation in expression of particular genes in heart-muscle cells in different parts of the heart with differences in the electrical properties of the individual cells. These can then be included in models of the ventricular wall that can reproduce the dynamics seen in ECG recordings. As this illustrates, the combined powers of genetics, biochemistry and electrophysiology begin to produce a rich picture of how all the millions of beating heart-muscle cells can work together as a co-ordinated system, and of what can go wrong with it. This kind of background understanding of the heart will help researchers plan development of cell therapies, and gene therapies, for hearts. There are already promising results in animals from work with artificial cell transport proteins, which can help correct problems with regulation of heart rhythm.

If these cellular or genetic strategies for repair do not quite do the trick, there will always be surgery, and modern techniques are making it easier to get at the heart, and allowing finer work. In particular, 'robotic' surgery – usually a misnomer since it typically involves close control of remote tools by a human operator – allows some repair work to be done more delicately than traditional scalpel-wielding can manage. One application is in coronary bypass surgery, a serious procedure involving large incisions, stopping the heart and keeping the patient going on a heart-lung machine. The incisions to get at the heart are becoming smaller, but robot-assisted operations can use even finer openings. Robotic arms can reach between the ribs so the breastbone does not need to be opened up, and they cut their way into the tough outer covering of the heart, the pericardium, and do the arterial replumbing required. Robot-assisted surgery has also been used to attach leads to the left ventricle to allow a pacemaker to resynchronise the ventricles when irregularities develop after heart failure, and even for the extremely delicate surgery of heart valves.

Replacing the Heart

If all this tinkering or repair of the wetware of human tissue – blood vessels, muscle fibres, even electrical conduction channels – will still not get the pump back in working order, why not replace the whole thing? The heart transplant was established as a last resort in the final decades of the twentieth century. It is a last resort partly because of the scale of the surgery involved and the cost of drugs to prevent rejection afterwards, partly because every recipient relies on a healthy human dying with their heart in good repair. Since usable human hearts are in perennially short supply (car accidents, that other self-inflicted epidemic, just do not occur often enough to clear the transplant queue), there has been a good deal of work recently on making other animals suitable donors for humans. This has been a recurrent fantasy since the flamboyant Frenchman Alexis Carrel's first attempts at transplant surgery in the early twentieth century in the USA. It faltered for the same reason that human-to-human transplants were long impossible – what Primo Levi called the savage mistrust that every living structure harbours toward any material of living origin. In other words, the immune system.

In normal transplants, this mistrust is now overcome with immunosuppressive drugs. Try an interspecies transplant, though, and the differences are too great for this to work. The new organ is quickly reduced to a blackened, necrotic mass. But, in one of the more ingenious applications of our new biological virtuosity, genetic engineering can now alter the cell-surface antigens to fool the recipient's immune system into treating an incoming piece of tissue as one of their own.

In principle, this means that organs from the animal of choice can now be made ready for human use. That animal? The pig. It is roughly the same body mass as an adult human, and its heart even beats at more or less the same rate. Heart valves from pigs have been used in human surgery for some years. Pig hearts transplanted into baboons have worked for a few months, although gradually worsening blood clots have done for the recipients in the end.

The bugbear of this work, though, has been the possibility that pig tissue has hidden within it gene sequences from viruses – 'porcine endogenous retroviruses' or PERVs – that might welcome the chance to make mischief in a new environment. Our recent experience with retroviruses that appear to have arisen in other species (HIV is one such) has induced extreme caution, and any move to clinical trials of xenotransplantation is currently stalled.

However, knowledge of the precise make-up of whole genomes, and ability to scan new ones, is developing so fast that sequence-gazers may one day be able to rule out any chance of genetic contamination from a pig-heart donor. Will we then see farms of healthy, and presumably well-tended, porkers destined for the operating theatre rather than the slaughterhouse? A few companies are already planning on this – and foresee billion-dollar rewards from their high-tech pigsties.

Meantime, there is another, non-biological route to substitution. Maybe we can use regular, metal and plastic technology to engineer our way out of heart trouble. After all, the thing *is* a pump, and we had already built pumps of our own to inspire Harvey's insights into the circulation 400 years ago. Now we have discovered a pump inside ourselves, isn't it time we built a replacement?

In principle, of course, it can be done. Open-heart surgery, let alone transplants, would not be possible without a mechanical pump that can take over the patient's circulation while they are on the operating table. And we have plenty of experience with replacing small components – heart valves again – with artificial versions, and even with implanting emergency control units. Tens of thousands of pacemakers are fitted in the US every year.

The specification for an artificial heart that actually sits inside your chest for the rest of your life is exceptionally demanding, though. It needs a power source, a control

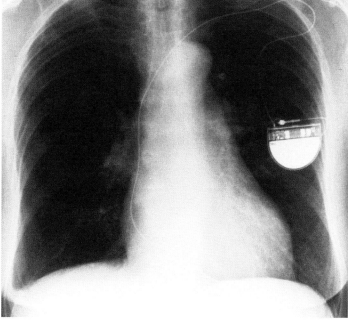

4a Cardiac pacemaker

4b X-ray of pacemaker in situ

system, and utter reliability for billions of beats. More subtly, it has to be made of materials that do not provoke inflammation or other cell damage, which can lead to arterial blockage that defeats the whole object of the exercise.

All this means that the history of artificial hearts is inglorious, with a whiff of headline-chasing taking precedence over genuine prospects of helping patients. Admittedly, there comes a point where you have to try the thing for real, and a dying patient may be happy to donate their body to research a little ahead of time. In any case, the various technologies in play are now maturing into usable, albeit still largely experimental, options.

A recent review reckons that there are thirty different 'mechanical circulatory devices' either in use or in preclinical trials. This more general term covers gizmos that help an ailing heart – such as the 'Heartmate left ventricular assist device', which does what it says on the tin. (The left ventricle is the target because that is the one that does most of the work, impelling the blood round the whole body where the right ventricle only has to send it round the lungs and back.) It was originally used to keep people going until they could have a transplant, but has now been approved in the US as a 'destination therapy', meaning that's all you get.

But devices that are actual heart replacements are looking more promising, too. In September 2006, the Food and Drug Administration in the US approved the AbioCor®

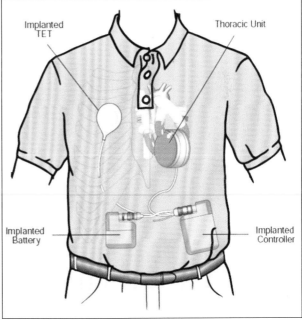

5a AbioCor® implantable replacement heart   **5b** Diagram of AbioCor® in situ

implantable replacement heart for use in patients who are in end-stage heart failure. That is, they will die unless they are allowed a go with this latest mechanical replacement. The makers have not quite perfected the specification yet – the current model is too big to fit in women, and even a male patient needs quite a large chest cavity. But it is a marvel of sophistication compared with the artificial hearts of the 1980s. In particular, it is powered by external transmitter and internal receiver coils, which charge its on-board battery, so is the first artificial heart with no wires connecting it to the outside world. The patient still has to stay close to an array of equipment for power supply, diagnostics and computer control, but can walk around and even leave the hospital. The manufacturers, Abiomed, are developing a new model that is a third smaller, and could work for 'up to five years'. This is some way short of a lifetime, but a huge bonus for people who are dying of heart failure.

For some designers, though, the AbioCor® device may be too closely modelled on the normal heart. Ventricular-assist gadgets usually have continuous flow pumps, driven by a simple rotor. Why not use the same kind of thing to replace the entire heart? They could be much smaller and, in theory, more hard-wearing. A team at the Texas Heart Institute in Houston has already tried a continuous-flow heart in calves, although only for a matter of weeks. But they see no reason why the technology cannot be developed to produce a reliable device that would match, or even improve on, the human heart. The team has even speculated that the mid-twenty-first-century Olympics may need special awards for athletes with superior mechanical hearts, who will no longer compete with sportsmen and women who grew their own. Just one thing, though. This model, perhaps the ultimate technical fix, *is* rotary. There is no regular muscular squeeze. No heartbeat. Now that really would bring a new sensation to the users. It would be not merely a redesign of a component, but of a fundamental human experience. Users may have Olympian fitness, and their blood may move, but they will have no pulse.

About the Contributors

Dr Ted Bianco is Director of Technology Transfer at the Wellcome Trust. His background is in tropical diseases and most of his research career has been devoted to the study of parasitic nematodes – ironically, a class of organisms that do not have a heart. Having spent twenty-five years studying pathogenic organisms, he now works to facilitate the development of new treatments for infectious and non-communicable diseases.

Dr Fay Bound Alberti is Lecturer in History at the University of Lancaster. She has published on aspects of the history of medicine, emotions and subjectivity between 1600 and 1900, and is the editor of *Medicine, Emotion and Disease, 1700–1950* (2006). She is currently preparing a monograph entitled *The Heart of the Matter: Locating Emotions in Medical and Cultural History*.

Michael Bracewell is the author of six novels and two works of non-fiction, including a selection of journalism and essays, *The Nineties*. His work has been included in many anthologies, including *The Penguin Book of Twentieth Century Fashion Writing* and *The Faber Book of Pop*. He is a regular contributor to *frieze* magazine, and has written catalogue essays on many contemporary artists.

Melissa Larner is a freelance editor and writer. She has edited numerous books, catalogues and articles on the arts and culture for Tate Publications, Phaidon Press, the Serpentine Gallery and *frieze* magazine, among many others.

Sir Jonathan Miller is a physician, a director of theatre and opera, and an author and lecturer. He has written and presented several major series for the BBC. His books include *The Human Body, The Facts of Life, On Reflection* and *The Body in Question*, based on his thirteen-part TV series on the history of medicine.

Dr Ayesha Nathoo is a post-doctoral teaching associate at King's College, University of Cambridge, and an affiliated research scholar of the Department of History and

Philosophy of Science. Her forthcoming book *Hearts Exposed: Transplants and the Media in 1960s Britain* explores the cultural history of heart transplantation and medical-media relations in the 1960s.

James Peto is Curator of Public Programmes at the Wellcome Trust and curator of the exhibition *The Heart* at Wellcome Collection, summer 2007. Previously he was Exhibitions Curator at the Whitechapel Art Gallery and Head of Exhibitions at the Design Museum, London.

Emily Jo Sargent is an artist and curator based in London. She has previously worked on several multidisciplinary exhibitions for the Wellcome Trust at the Science Museum and is co-curator of *The Heart.*

Jon Turney is a science writer and editor, and course leader for the MSc in Creative Non-fiction at Imperial College, London. He has been Features Editor for *The Times Higher Education Supplement,* head of Science and Technology Studies at UCL and an Editorial Director for Penguin Press. His books include *Frankenstein's Footsteps: Science, Genetics and Popular Culture* (1998), *Lovelock and Gaia: Signs of Life* (2004) and, with Dr Jess Buxton, *The Rough Guide to Genes and Cloning* (2007).

Dr Francis Wells is Consultant Cardiothoracic Surgeon at Papworth Hospital in Cambridge, and Associate Lecturer in the Department of Surgery at the University of Cambridge. He is also a Special Consultant on Leonardo's anatomical drawings for the Universal Leonardo Project.

Sir Magdi Yacoub, FRS is an eminent heart surgeon and pioneer of techniques in heart and heart-lung transplantation. He is Professor of Cardiothoracic Surgery at Imperial College London, Director of Research at The Magdi Yacoub Institute, and founder and President of The Chain of Hope UK charity.

Louisa Young is the author of seven novels, a biography and *The Book of the Heart* (2002), a cultural history of this most peculiar organ. She also writes children's fiction, with her daughter, under the joint *nom de plume* Zizou Corder. She has a degree in history from Trinity College, Cambridge, is published in thirty-six languages, and has been listed for the Orange Prize.

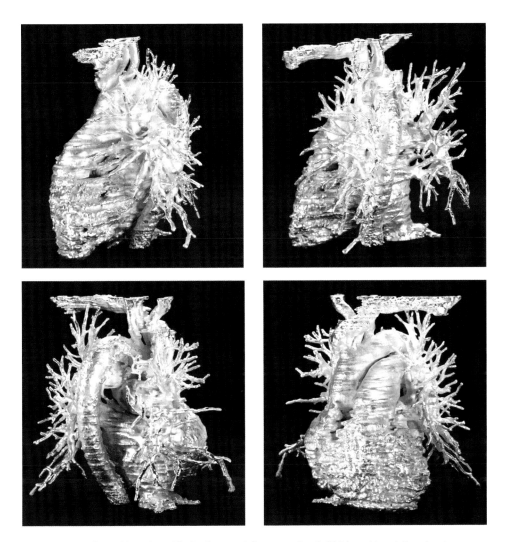

Jane Prophet, *Silver Heart* from 'Distinctions and Counterpoises', 2004, rapid prototyped polymer heart made from MRI data and subsequently copper and silver plated

Bibliography

Chapter 1

Davies, Nigel, *Human Sacrifice in History and Today*, Dorset Press, New York, 2000

Doueihi, Milad, *A Perverse History of the Human Heart*, Harvard University Press, Cambridge, MA, 1998

Erickson, RA, *The Language of the Heart 1660–1750*, University of Pennsylvania Press, Philadelphia, 1997

Grillot de Givry, Emile, *Witchcraft, Magic and Alchemy*, Dover Publications, New York, 1971

Jager, Eric, *The Book of the Heart*, University of Chicago Press, Chicago, 2000

Makes, Jakemes, *Le Roman du Châtelain de Couçi et la Dame de Fayel*, John E Matze and Maurice Delboville (eds), Société des Anciens Textes Français, Paris, 1936

Sahagún, Bernardino de, *The Florentine Codex*, Book 2, Arthur JO Anderson and Charles Dibble (ed & trans), University of Utah Press, Utah, 1981

Regnier-Bohler, Danielle (ed & trans), 'Le Lay d'Ignauré', in *Le Coeur Mangé: recits érotiques et courtois, xii et xiii siècles*, Stock, Paris, 1979

Wallis Budge, E, *The Hieroglyphic Vocabulary to the Book of the Dead*, Dover Publications, New York, 1991

Chapter 2

Cannon, Walter B, *Wisdom of the Body*, Peter Smith, New York, 1963

Gregory, Andrew, *Harvey's Heart: The Discovery of the Blood Circulation*, Icon, Cambridge, 2001

Harvey, William, *De Motu Cordis* (1628), Kenneth J Franklin (trans), in *Movement of the Heart and Blood in Animals*, Blackwell, London, 1957

Hippocrates, 'On the Heart', IM Lonie (trans), in GER Lloyd (ed), *Hippocratic Writings*, Penguin, London, 1978

Miller, Jonathan, *The Body in Question*, Jonathan Cape, London, 1978

Porter, Roy and Bynum, W (eds), *Companion Encyclopaedia of the History of Medicine*, Routledge, London, 1993

Whitteridge, Gweneth, *William Harvey and the Circulation of the Blood*, Macdonald/Elsevier, London, 1971

Young, JZ, *Introduction to the Study of Man*, Oxford University Press, Oxford, 1971

Chapter 3

Jacobs, Fredrika H, *The Living Image in Renaissance Art*, Cambridge University Press, Cambridge, 2005

Keele, KD, *Leonardo da Vinci on Movement of the Heart and Blood*, Harvey & Blythe, London, 1952

Kemp, Martin, *Leonardo da Vinci. The Marvellous Works of Nature and Man*, Oxford University Press, Oxford, 2006

O'Malley, Charles and Saunders, JB, *Leonardo da Vinci on the Human Body*, Dover Publications, New York, 1983

Rifkin, Benjamin, Ackerman, Michael and Folkenberg, Judith, *Human Anatomy: Depicting the Body from the Renaissance to Today*, Thames & Hudson, London, 2006

Young, Louisa, *The Book of the Heart*, Flamingo, London, 2002

Chapter 4

Blessed Raymond of Capua, *The Life of St Catherine of Siena*, George Lamb (trans), Harvill Press, London, 1960

Cassavas, Denys, *A Manual of Affective and Practical Devotion to the Sacred Heart of Jesus, from the writings of Blessed Margaret Mary Alacoque*, London, 1873

McLean, Adam (ed), *The Rosicrucian Emblems of Daniel Cramer*, Phanes Press, Grand Rapids, MI, 1991

Manning Stevens, Scott, 'Sacred Heart and Secular Brain', in David Hillman and Carla Mazzio (eds), *The Body in Parts: Fantasies of Corporeality in Early Modern Europe*, Routledge, London, 1997

Pears, E Allison (ed & trans), *The Life of Teresa of Jesus: The Autobiography of St Teresa of Avila*, Image Books, New York, 1991

St Bonaventura, 'The Perfect Life', in Sir John Robert O'Connell, *Devotion to the Sacred Heart of Jesus: its History and Practices* , Catholic Truth Society of Ireland, Dublin, 1940

Villeneuve di Gergy, Jean Joseph Languet de la, *The Life of the Venerable Mother Magaret Mary Alacoque*, Faber, London,1847

The Image of Christ (catalogue), National Gallery, London, 2000

Chapter 5

Bound Alberti, Fay, *The Heart and the Matter: Locating Emotions in Medical and Cultural History* (forthcoming, Oxford University Press)

Bound Alberti, Fay (ed), *Medicine, Emotion and Disease, 1700–1950*, Palgrave Macmillan, Basingstoke, 2006

Erickson, Robert A, *The Language of the Heart 1600–1750*, University of Pennsylvania Press, Philadelphia, 1997

Fuchs, Thomas, *The Mechanization of the Heart*, Boydell and Brewer, London, 2000

Galen, *On the Passions and Errors of the Soul*, PW Harkins (trans), Ohio State University Press, Columbus, 1963

Rather, Lelland J, 'Old and New Views of the Emotions and Bodily Changes: Wright and Harvery versus Descartes, James and Cannon', in *Clio Medica*, 1, 1965

Temkin, Owsei, *Galenism: Rise and Decline of a Medical Philosophy*, Cornell University Press, Ithaca, New York and London, 1973

Wright, Thomas, *Passions of the Minde in Generall* (1601, 1604), University of Illinois Press, Urbana, 1971

Chapter 6

Acierno, LJ, *The History of Cardiology*, Parthenon Publishing, New York, 1994

Bing, RJ, *Cardiology: The Evolution of the Science and the Art*, Rutgers University Press, New Brunswick, NJ, 1999

Dong, E, Shumway, N and Lower, R, 'A heart transplantation narrative: The earliest years', in P Terasaki (ed), *History of transplantation: Thirty-five recollections*, UCLA Tissue Typing Laboratory, Los Angeles, 1991

Fox, R and Swazey, J, *Spare parts: Organ replacement in American society*, Oxford University Press, New York, 1992

Lawrence, C, 'The "New Cardiology" in Britain 1880–1930', in WF Bynum, C Lawrence and V Nutton (eds), *The Emergence of Modern Cardiology*, Wellcome Institute for the History of Medicine, London, 1985

Leach, ER, *A Runaway World?: The Reith Lectures 1967*, British Broadcasting Corporation, London, 1968

Lederer, SE, 'Tucker's heart: Racial politics and heart transplantation in America', in K Wailoo, J Livingston and P Guarnaccia (eds), *A Death Retold: Jesica Santillan, the Bungled Transplant, and Paradoxes of Medical Citizenship*, University of North Carolina Press, Chapel Hill, 2006

Lock, M, *Twice Dead: Organ Transplants and the Reinvention of Death*, University of California Press, Berkeley and London, 2002

McKellar, S, 'Artificial Hearts: A Technological Fix More Monstrous than Miraculous?', in L Rosner (ed), *The Technological Fix: How people use technology to create and solve problems*, Routledge, New York and London, 2004

McRae, D, *Every Second Counts: The Race to Transplant the First Human Heart*, Simon & Schuster, London, 2006

Nathoo, A, *Hearts Exposed: Transplants and the Media in 1960s Britain*, Palgrave Macmillan, London, 2007

Richardson, R, *The Surgeon's Heart: A History of Cardiac Surgery*, Heinemann Medical, London, 1969

Silverman, ME, Fleming, PR, et al (eds), *British Cardiology in the Twentieth Century*, Springer-Verlag, London, 2000

Sylvia, Claire, *A Change of Heart*, Warner Books, Clayton, Victoria, 1998

Tansey, T and Reynolds, L (eds), *Early Heart Transplant Surgery in the UK: The Transcript of a Witness Seminar Held at the Wellcome Institute for the History of Medicine, London, on 10 June 1997*, Wellcome Trust, London, 1999

Weisse, Allen B, *Heart to Heart: The Twentieth Century Battle Against Cardiac Disease*, Rutgers University Press, New Brunswick, NJ, 2002

Westaby, S, *Landmarks in Cardiac Surgery*, Isis Medical Media, Oxford, 1997

Chapter 7

Adam J, 'Making hearts beat', *Innovative Lives*, Smithsonian Institution biography of Wilson Greatbatch, 1999, *http://invention.smithsonian.org/centerpieces/ilives/lecture09.html*

Blackstone E, Morrison M and Roth MB, 'H2S induces a suspended animation-like state in mice', *Science*, 308, 2005

Hill, Richard, Wyse, Gordon and Anderson, Margaret, *Animal Physiology*, Sinauer Associates, Sunderland, MA, 2004

Randall, David, Burggren, Warren and French, Kathleen, *Eckert Animal Physiology*, WH Freeman, New York, 2001

Schmidt-Nielsen, Knut, *Animal Physiology: Adaptation and Environment*, Cambridge University Press, Cambridge, MA, 1997

Chapter 8

Goldman, Albert, *Elvis*, McGraw-Hill, New York, 1984

Leviton, Jay B, *Elvis Close-Up*, Century Hutchinson, London, 1989

Marcus, Greil, *Dead Elvis: Chronicle of Cultural Obsession*, Doubleday, New York, 1991

Savage, Jon, *Time Travel*, Chatto & Windus, London, 1996

Chapter 9

Gray, Norman and Selzman, Craig, 'Current Status of the Totally Artificial Heart', *American Heart Journal*, 152, 2006

Jeffrey, K, *Machines in our hearts: The cardiac pacemaker, the implantable defibrillator, and American health care*, Johns Hopkins University Press, Baltimore, MD and London, 2001

Mathur, A and Martin, J, 'Stem Cells and Repair of the Heart', *The Lancet*, 364, 2004

Noble, Denis, 'Modeling the Heart – From Genes to Cells to the Whole Organ', *Science*, 1 March 2002

Olsen, Eric, 'Gene Regulatory Networks in the Evolution and Development of the Heart', *Science*, 29 September 2006

Poole-Wilson, Philip, 'The Heart and Circulation', in Marshall Marinker and Michael Peckham (eds), *Clinical Futures*, BMJ Books, London, 1998

Swedberg, Karl, 'The Future – Cardiovascular Medicine in 10 Years', *Heart*, 84, 2000

Index

Page numbers in italics refer to illustrations